Organic photochemistry

Cambridge Texts in Chemistry and Biochemistry

Organic photochemistry

J.M.COXON
University of Canterbury, NZ

B.HALTON
Victoria University of Wellington, NZ

SECOND EDITION

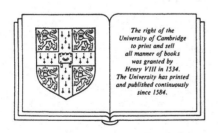

The right of the
University of Cambridge
to print and sell
all manner of books
was granted by
Henry VIII in 1534.
The University has printed
and published continuously
since 1584.

CAMBRIDGE UNIVERSITY PRESS
Cambridge
London   New York   New Rochelle
Melbourne   Sydney

CAMBRIDGE UNIVERSITY PRESS
Cambridge, New York, Melbourne, Madrid, Cape Town, Singapore,
São Paulo, Delhi, Dubai, Tokyo, Mexico City

Cambridge University Press
The Edinburgh Building, Cambridge CB2 8RU, UK

Published in the United States of America by Cambridge University Press, New York

www.cambridge.org
Information on this title: www.cambridge.org/9780521189729

First published 1974
Second edition 1986
First paperback edition 2011

A catalogue record for this publication is available from the British Library

Library of Congress Cataloguing in Publication data

Coxon, J. M. (James Morriss), 1941–
  Organic photochemistry.

  (Cambridge texts in chemistry and biochemistry)
  Includes bibliographies and index.
  1. Photochemistry. 2. Chemistry, Organic.
1. Halton, B. (Brian) II. Title. III. Series.
QD708.2.096 1986 547.1'35 85-24251

ISBN 978-0-521-32067-2 Hardback
ISBN 978-0-521-18972-9 Paperback

CONTENTS

PREFACE TO THE FIRST EDITION

The use of light to effect chemical change has been recognised for many years, but it is only recently that sufficient knowledge has been attained to place photochemical reactions in the realm of organic synthesis. The recent application, by Woodward and Hoffmann, of the principle of conservation of orbital symmetry to concerted reactions has made an important contribution to the understanding of many photochemical processes. This book has been written to provide an introduction to the principles and applications of organic photochemistry at a level suitable for senior undergraduate and graduate students in universities and technical institutes. It is not intended to provide an exhaustive survey of the field but rather to provide the student with an up-to-date background of the subject, on which a more detailed study can be based.

The authors gratefully acknowledge many helpful comments from Dr K. Schofield. We also thank Dr B. G. Odell for critically reading the entire manuscript, and Professors M. F. Grundon and J. Vaughan, and Dr M. P. Halton and Mr A. D. Woolhouse for many helpful suggestions. Any errors are the sole responsibility of the authors. Finally we thank our wives.

J. M. C.
B. H.

New Zealand, 1972

PREFACE TO THE SECOND EDITION

In the decade since this book first appeared research involving organic photochemistry has been prolific. In this edition we have attempted to summarise those classes of reaction which best illustrate the types of photochemical behaviour commonly observed for simple organic molecules. Wherever possible reference is given to review-type material for the student or teacher wishing to pursue the topic in more detail; the annual Royal Society of Chemistry specialist periodical report *Photochemistry* provides an excellent route to the primary literature for those who seek such detail.

We anticipate that the use of lasers to investigate photochemically induced reactions will become more common in the next ten years. Thus much more detailed information on known reactions and of specific excited states and their chemistry is likely to become available.

The authors acknowledge many helpful comments from Professor K. Schofield. We also thank Dr P. J. Steel and Dr M. P. Halton for reading the manuscript and for their constructive suggestions. We also appreciate the many helpful and encouraging comments from colleagues around the world who have used the first edition for their courses.

J. M. C.

New Zealand, 1986   B. H.

1 Introduction – excitation and the excited state

Photochemistry is the study of chemical reactions initiated by light. The interaction of electromagnetic radiation with matter covers a very wide field but it is only recently that this area of chemistry has been given serious and systematic attention. The quantum mechanical theory, developed in the late nineteen twenties, has helped in particular to rationalise the interaction of light with matter. Despite this major advance in theory, the study of organic photochemistry did not progress significantly because of the lack of suitable ultraviolet light sources and the difficulty in analysing unusual, and often complex, product mixtures. Improvements in analytical techniques, the development of spectroscopic methods in organic chemistry, and the availability of commercial ultraviolet light sources, have all contributed to the rapid expansion of organic photochemistry.

1.1 The interaction of electromagnetic radiation with matter[1-4]

Electromagnetic radiation can be regarded as having a dual nature. It is propagated through space, in wave form, obeying the relationship $c = v\lambda$, where c is the speed of electromagnetic radiation $(2.9979 \times 10^8 \text{ m s}^{-1})$, v the frequency and λ the wavelength of the radiation. At the same time, the absorption or emission of radiation by matter occurs only in discrete quanta (photons) and is governed by the relationship

$$E = hv = hc/\lambda$$

where E is the energy absorbed or emitted and h is the constant of proportionality or Planck's constant $(6.6256 \times 10^{-34} \text{ J s})$. The amount of energy absorbed or emitted is inversely proportional to the wavelength of the radiation – short wavelengths correspond to high energy absorption and long wavelengths to low energy absorption.

When a molecule absorbs energy the process is referred to as *excitation*. The molecule is raised from its *ground state* of minimum energy, to an *excited state* of higher energy. In the infrared, visible, or ultraviolet regions of the electromagnetic spectrum this may involve excitation in the rotational, vibrational or electronic energy levels. A change in the rotational levels of a molecule is characterised by low energy (long wavelength) absorption in the far infrared, while an increase in the vibrational energy of a molecule requires 10–100 times more energy and occurs in the infrared region of the spectrum. An increase in molecular energy by excitation within the electronic levels requires promotion of an electron from one molecular orbital to another of higher energy, and this is an even more energetic process than vibrational excitation. In fact energy in excess of 10 times that required for vibrational excitation is required, with absorption occurring in the ultraviolet and visible regions of the spectrum. It is the last process (electronic excitation) that is of prime importance in organic photochemistry, although it is inevitably accompanied by some increase in vibrational and rotational energy.

Absorption by a molecule, of radiation in the ultraviolet (200–400 nm) or visible (400–800 nm) region of the spectrum can result in an excited state so high in energy that the energy absorbed is comparable in magnitude with the bond dissociation energies associated with organic molecules. For one molecule the energy of excitation $= hc/\lambda$ and for one mole the energy of excitation $= Lhc/\lambda$, where L is the Avogadro constant,

$$= \frac{6.0225 \times 10^{23} \times 6.6256 \times 10^{-34} \times 2.9979 \times 10^{8}}{\lambda(\text{nm}) \times 10^{-9}} \, \text{J mol}^{-1}$$

$$= \frac{1.20 \times 10^{5}}{\lambda(\text{nm})} \, \text{kJ mol}^{-1}$$

If absorption occurs at 250 nm, then the energy associated with this transition ($E = 480 \, \text{kJ mol}^{-1}$) is greater than the bond dissociation energy of a carbon–carbon σ-bond ($D \sim 347 \, \text{kJ mol}^{-1}$). It is not surprising, therefore, that chemical reaction can be induced by excitation with ultraviolet light. The first law of photochemistry, the Grotthus–Draper law, states that only the radiation absorbed by a molecule can be effective in causing chemical change. However, not every photon ($h\nu$) absorbed by a molecule will necessarily produce a chemical change. The excitation energy can be lost by fluorescence or phosphorescence (p. 7), or by molecular collision.

1.2   Excitation[1-4]

Interaction of matter with electromagnetic radiation results in excitation of the matter to a higher energy level. With electronic excitation the transition that the electron undergoes can be classified according to the orbitals involved.

The Schrödinger wave equation describes the electron distribution in the hydrogen atom by the atomic orbitals s, p, d and f and these orbitals can be used for all other atoms. Molecular orbitals are formed by a combination of the atomic orbitals, the most common being a *linear combination of atomic orbitals* (LCAO). A given number of atomic orbitals always gives rise to the same number of molecular orbitals. Combination of two atomic orbitals, on different atoms, by overlap along the internuclear axis results in the formation of two molecular orbitals, a bonding orbital (either σ or π), and an antibonding orbital (σ^* or π^*). The overlap of spherical s-orbitals always results in the formation of the molecular orbitals σ and σ^*, whereas the molecular orbitals formed by the linear combination of two p-orbitals will depend on their relative orientation. End-on overlap, along the internuclear axis, leads to σ- and σ^*-molecular orbitals (fig. 1.1(*a*)) while side-on overlap results in a bonding π- and an antibonding π^*-molecular orbital (fig. 1.1(*b*)). The wave function used to describe an atomic p-orbital shows the two lobes to have opposite phases, one mathematically positive and the other mathematically negative. Throughout this text this phase difference will be depicted by shading one of the lobes as illustrated in fig. 1.1. A bonding combination of atomic orbitals requires the overlap of orbitals of like phase (sign) and is characterised by a concentration of electron density between the bonding atoms. Overlap of orbitals of unlike phase (sign) leads to an anti-bonding molecular orbital characterised by a depletion of electron

Fig. 1.1. The formation of molecular orbitals from atomic orbitals.

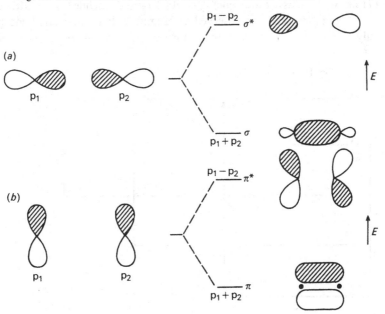

density between the bonding atoms (fig. 1.1). Orbitals not involved in the LCAO process are described as non-bonding n-orbitals, the energies of which are generally between that of the highest bonding and lowest antibonding molecular orbital.

Two electrons are assigned to each molecular orbital such that their spins are paired and, in general, a molecule in its ground state has all of its electrons spin-paired in the bonding (and non-bonding) orbitals. Electronic excitation of a molecule results in the promotion of an electron from one molecular orbital to another of higher energy, e.g. σ–σ^*, n–σ^*, π–π^* and n–π^*. In simple molecules excitation of an electron from the σ- to the σ^*-orbital requires the greatest energy (fig. 1.2) and the wavelength of the absorption associated with this transition is generally below 150 nm in the far ultraviolet (10–200 nm), a region of the electromagnetic spectrum only now becoming accessible with the advent of tunable lasers. This transition is therefore of limited significance to the organic photochemist and this is also true of the n–σ^* transition, requiring wavelengths of excitation of c. 200 nm. It is the alternative excitation processes (π–π^* and n–π^*) that are responsible for the bulk of organic photochemical reactions. The wavelength of light associated with these two excitations is observed in accessible regions of the ultraviolet and visible part of the electromagnetic spectrum.

The absorption wavelength for any particular transition is dependent on the structure of the molecule. For ethylene the π–π^* transition occurs at 171 nm with an associated energy of 700 kJ mol^{-1}, while for butadiene the lowest π–π^* transition is observed at 214 nm with an associated energy of only 560 kJ mol^{-1}. This latter transition occurs by promotion of an

Fig. 1.2. Electronic excitation processes.

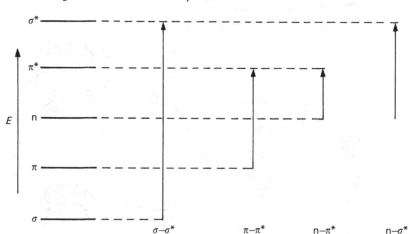

electron from the *highest occupied molecular orbital* (HOMO) to the *lowest unoccupied molecular orbital* (LUMO) as shown in fig. 1.3 where the π-molecular orbitals of the diene are depicted by atomic p-orbitals which are presumed to overlap. Throughout this text molecular orbitals will be represented as overlapping atomic orbitals for clarity. As conjugation is increased the energy required for $\pi-\pi^*$ excitation decreases and the transition occurs at longer wavelength. The $n-\pi^*$ excitation of saturated carbonyl compounds occurs at *c.* 280 nm and corresponds to an energy of *c.* 430 kJ mol^{-1}.

The $\pi-\pi^*$ excitation process (fig. 1.4) can result in two possible electron arrangements. The arrangement with the electron spins paired is termed the *singlet state*, and designated S_1, and that with the electron spins parallel is termed the *triplet state*, and designated T_1. This nomenclature follows from

Fig. 1.3. The π-orbitals of buta-1,3-diene showing the HOMO and the LUMO.

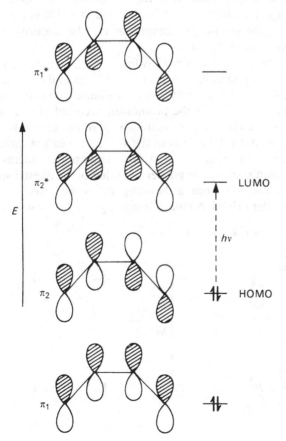

the multiplicity (M) observed in atomic absorption and emission spectra. M is defined as $2S + 1$, where S is the total spin of the system. Thus with a spin-paired system, $S = 0$ and $M = 1$ (singlet), and for a spin-parallel system, $S = 1$ and $M = 3$ (triplet). The subscript 1 after S and T defines the excited states as the first excited singlet and triplet respectively. The ground state with the electron spins paired is consequently a singlet and designated S_0. Of the excited states S_1 and T_1, the triplet is generally of lower energy in accord with Hund's rule, which states that the most stable arrangement of electrons in atoms (or molecules) is that with maximum multiplicity. In addition to this, the triplet will have a longer lifetime than the singlet state since a spin inversion must accompany any deactivation of the triplet to the ground state. For ethene however the first excited singlet state is lower in energy than the first excited triplet state. The n–π* excitation of a carbonyl group is depicted in fig. 1.5, and again singlet and triplet excited states are possible. The n–π* excitation is energetically more favoured than the π–π* excitation of the carbonyl group and represents the HOMO–LUMO transition. Although the latter process is observed at *c.* 170 nm, it is the former excitation process that is responsible for the majority of the photochemical reactions of ketones.

Electronic transitions occur very rapidly (*c.* 10^{-15} s), more rapidly than the time required for a molecular vibration (*c.* 10^{-13} s), and consequently the geometry of the excited state produced will be initially the same as that of the ground state from which the promotion occurred. The Franck–Condon principle states that the relative nuclear positions are unaltered in electronic excitation and corresponds to the vertical transition shown on the potential energy diagram (fig. 1.6). The equilibrium internuclear distance of the excited state will be greater than that of the ground state, a consequence of excitation from a bonding molecular orbital to an antibonding molecular orbital. A molecule in the ground state will exist in

Fig. 1.4. π–π* excitation of ethene.

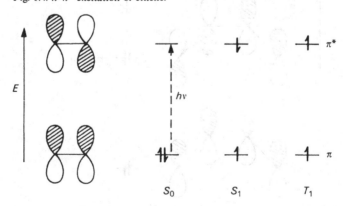

S_0 S_1 T_1

one of its lowest vibrational levels (represented by the horizontal lines in fig. 1.6) and excitation obeying the Franck–Condon principle will afford the first excited singlet of the molecule high in vibrational energy. The excess vibrational energy of the excited state will be rapidly lost by molecular collisions and dissipated as heat. Electronic transitions can occur from any vibrational level of the ground state to any vibrational level of the excited state. Thus the energy required to effect an electronic transition will vary within a limited range and give rise to a broad absorption band as observed in ultraviolet spectra.

Fig. 1.5. n–π* excitation of the carbonyl group.

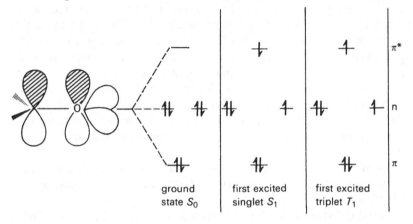

ground
state S_0

first excited
singlet S_1

first excited
triplet T_1

Fig. 1.6. Energy diagram showing excitation and radiative deactivation.

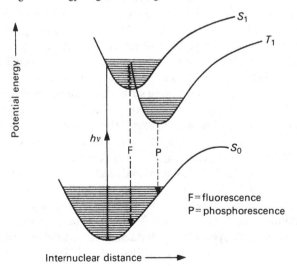

F = fluorescence
P = phosphorescence

The intensity of an absorption band follows two empirical laws: Lambert's law, which states that the fraction of radiation absorbed is independent of the source intensity, and Beer's law, which states that the radiation absorbed will be directly proportional to the solution concentration. While this latter does not hold over a wide concentration range, in dilute solutions the deviations are small. The following equation, known as the Beer–Lambert law, is a mathematical expression of these laws:

$$\log_{10} \frac{I_0}{I} = A = \varepsilon c l \qquad \text{(Beer–Lambert law)}$$

I_0 and I are the intensities of the incident and transmitted radiation respectively, A is defined as the absorbance (or optical density), c the concentration (mol dm^{-3}) and l is the path length (cm). The constant of proportionality ε is defined as the molar extinction coefficient and is a measure of a transition probability. For a singlet π–π^* excitation, ε normally has a value in the range 10^2–10^4 m^2 mol^{-1}, indicating a high probability of excitation, while for a singlet n–π^* process ε is usually in the range 1–5 m^2 mol^{-1}. For singlet to triplet excitation processes, the extinction coefficients are very small and less than unity, indicating the low probability of these transitions.

The application of quantum mechanical theory to electronic excitation processes has led to a set of *selection rules* by which transitions can be classified as allowed or forbidden. All excitation processes involving conservation of spin (or multiplicity) are allowed, e.g. S_0–S_1, T_1–T_2, while those involving a change in spin are *spin forbidden*, e.g. S_0–T_1. Excitation processes between states with the same symmetry are allowed, e.g. π–π^* singlet excitation of ethylene, while transitions between states of different symmetry are *symmetry forbidden*, e.g. n–π^*. Although S_0–T_1 excitation is a spin-forbidden process, it is, in fact, often observed owing to a breakdown in the selection rules. The selection rules result from calculations based on the electronic states being 'pure'. In practice the true nature of a given state, singlet or triplet, will include some mixing with other states, which is due to an interaction between the electron spin vector and the orbital angular momentum vector. This is known as spin–orbit coupling. Thus a first excited triplet state is more accurately described as largely T_1 with small components due to S_0, S_1, T_2 and perhaps other excited states. Mixing of states is enhanced by magnetic fields, unpaired electrons, and heavy atoms. For example, 9,10-dibromoanthracene has a more intense S_0–T_1 absorption than has anthracene itself. The low intensity of S_0–T_1 transitions preclude these from being an effective method of populating the triplet excited state. The symmetry-forbidden n–π^* singlet excitation is usually observed with an extinction coefficient of sufficient magnitude to obtain a significant excited singlet state population. Failure of the selection rules to hold rigorously for symmetry-forbidden transitions results from a modification of the ground (or excited) state symmetry by vibrational motion of the molecule. A vibrational mode has characteristic symmetry; it may differ from that of the stationary chromophore and when this is so any description of the ground (or excited) state symmetry must include the vibrational components. A transition can occur when any of the symmetry components are common to both the ground and excited states.

1.3   The excited state[1-8]

The shorter lifetime and higher energy of the singlet compared with the triplet excited state may be attributed to a difference in electron spin. The longer lifetime of an excited triplet species reflects the need for a spin inversion to accompany any deactivation. In general, singlet excited states have lifetimes in the range 10^{-9}–10^{-5} s while triplets are much longer lived and have lifetimes in the range 10^{-5}–10^{-3} s. An estimate of the singlet excited state lifetime τ, is obtained from the following expression

$$\tau \sim \frac{10^{-5}}{\varepsilon_{\text{max}}}$$

where ε_{max} is the molar extinction coefficient at the wavelength corresponding to maximum absorption. For example, the π–π^* singlet excitation of ethylene has a maximum of 171 nm with an extinction coefficient of 1553 $\text{m}^2 \text{mol}^{-1}$ resulting in an estimated singlet lifetime of 6.4×10^{-9} s, while the singlet n–π^* excitation of acetone has a maximum at 279 nm with an extinction coefficient of 1.5 $\text{m}^2 \text{mol}^{-1}$ resulting in an estimated singlet lifetime of 6.7×10^{-6} s.†

The difference in spin between excited singlet and triplet species gives rise to different properties. The singlet spin-paired state is diamagnetic while the triplet state, with two unpaired spin-parallel electrons, is paramagnetic. For any reaction to proceed from the triplet state, paramagnetic quenchers and free radical scavengers, such as molecular oxygen, must be excluded. The most powerful diagnostic tool for characterisation of a triplet state species is *electron paramagnetic resonance* spectroscopy (e.p.r.)[5]. E.p.r. is similar to nuclear magnetic resonance but is a consequence of *electron* spin rather than *nuclear* spin. Since all singlet states are spin-paired ($S = 0$), this method is not applicable to their detection. A triplet state, with a total spin $S = 1$, consists of a set of three sublevels arising from the magnetic quantum

---

† The wavelength of the π–π^* and n–π^* excitations are solvent-dependent. On increasing the solvent polarity, the π–π^* absorption generally moves to longer wavelength (a bathochromic shift) whereas the n–π^* band usually moves to shorter wavelength (a hypsochromic shift).

numbers, $M_S = 0, \pm 1$. In an external magnetic field these sublevels are of different energy and it is possible, by using the e.p.r. technique, to observe electron transitions between them. Initially applied to an examination of the naphthalene triplet, the e.p.r. technique has become increasingly important in both diagnosis and examination of the triplet state. From these observations the lifetime of the triplet species can be obtained. Laser spectroscopy is providing increasing amounts of information on short-lived excited states.[6]

The simple energy diagram (fig. 1.6) shows different equilibrium internuclear distances for ground, excited singlet and triplet states. The singlet and triplet state energies resulting from π–π* excitation are dependent upon the angle of twist about the carbon–carbon σ-bond. The excited states tend to stabilise themselves by distortion from the ground state geometry. For alkenes the energy minimum of the singlet and triplet

S_0 S_1 or T_1, E_{min}

excited states corresponds to the orbitals being orthogonal. In terms of the Franck–Condon principle excitation produces an excited state with planar geometry and distortion and rotation about the carbon–carbon bond occurs after the excitation step. The geometric changes produced by n–π* excitation of simple carbonyl compounds are quite pronounced. In the ground state the carbon–oxygen bond distance of formaldehyde is 0.121 nm whereas in the excited singlet state this is increased to 0.132 nm and the molecular geometry changes from planar to pyramidal with an out-of-plane angle of c. 25°. In the first excited triplet state the out-of-plane angle is further increased to c. 35° although the carbon–carbon bond distance is not significantly affected.

S_0 planar S_1 pyramidal T_1 pyramidal

Once formed, the excited singlet and triplet states will either undergo chemical reaction or lose their excitation energy by a radiative or non-radiative process (fig. 1.7). There are two types of radiative deactivation, *fluorescence* and *phosphorescence*. Fluorescence is the emission of radiation accompanying the deactivation of an excited species to a lower state of the same multiplicity, e.g. S_1–S_0, T_2–T_1, and is a spin-allowed process. Phosphorescence is the emission of radiation accompanying the deactivation of an excited species to a lower state of different multiplicity, e.g. T_1–S_0, and is a spin-forbidden process. Excitation results in the occupation of excited states high in vibrational energy. This energy is rapidly lost (*c.* 10^{-10} s) by collisional transfer, a process referred to as *vibrational cascade*, and radiative deactivation normally occurs from the lowest vibrational level of the excited state. As a consequence it is usual for less energy to be emitted than is absorbed and the emission spectrum will be shifted to longer wavelength compared with the absorption spectrum (Stokes' law). Occasionally the excited state will gain thermal energy and move to a higher vibrational level prior to deactivation and the emission spectrum will be moved to a shorter wavelength than the absorption spectrum. This is termed anti-Stokes behaviour.

The non-radiative processes are of two general types, namely *internal conversion* (IC) and *intersystem crossing* (ISC). IC involves the transition from one state to another of the same multiplicity without loss of energy. This is demonstrated by the S_1–S_0 process (fig. 1.7) where the ground state is initially formed in a high vibrational level and subsequently undergoes rapid vibrational cascade to its lowest energy level. ISC is the conversion of one state to another of different multipicity without loss of energy and provides the most favourable route to the triplet state. At the point in the energy diagram (fig. 1.6) where the excited singlet and triplet cross, the potential and kinetic energies of both states are the same and ISC from the singlet to the triplet state can occur. Clearly ISC is most efficient when the singlet and triplet excited states are of comparable energy. The triplet so produced will be high in vibrational energy and vibrational cascade will

Fig. 1.7. A Jablonski diagram showing excitation and deactivation routes.

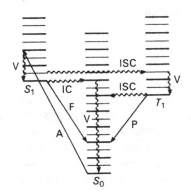

A = absorption ($\sim 10^{-15}$ s)
F = fluorescence (10^{-9}–10^{-5} s)
P = phosphorescence (10^{-5}–10^{-3} s)
V = vibrational cascade ($\sim 10^{-10}$ s)
IC = internal conversion ($\sim 10^{-10}$ s)
ISC = intersystem crossing ($\sim 10^{-6}$ s)

rapidly follow. The reverse process of ISC, T_1–S_1, is highly unlikely since the most stable form of the triplet is lower in energy than the singlet state. The ease and efficiency of ISC varies from compound to compound, being dependent on the structural environment of the chromophore. In general the longer lived an excited singlet, the more liable it is to undergo ISC. Carbonyl compounds give a high triplet state population by this route.

Irradiation of a molecule normally results in the absorption of a single photon with subsequent competition between radiative and non-radiative deactivation, and chemical reaction. Not every photon absorbed will necessarily be effective in bringing about chemical change and the efficiency of a photochemical process is therefore of fundamental value. The efficiency of a photochemical process is defined by the product quantum yield Φ, where

$$\Phi = \frac{\text{the number of molecules reacting per unit volume per unit time}}{\text{the number of photons absorbed per unit volume per unit time}}$$

The magnitude of the quantum yield and the effects on it of variables, such as temperature, pressure and concentration, are of importance since they can provide valuable information as to the nature of the reaction. The quantum yield for formation of any specific product may or may not be the same as the quantum yield for decomposition of starting material, depending on the nature and complexity of the reaction. In most photochemical reactions the quantum yield for any particular product will range from zero to unity. However, in chain reactions, where the absorption of a photon initiates the reaction, the value of the quantum yield may be several powers of ten. For example, the free radical halogenation of alkanes proceeds by initial photolytic dissociation of the halogen molecule, with radical propagation steps reproducing halogen radicals. The quantum yield for this reaction is correspondingly large (c. 10^5).

$$Br_2 \xrightarrow{h\nu} 2Br\cdot \qquad \text{initiation}$$

$$\left. \begin{array}{l} RH + Br\cdot \longrightarrow R\cdot + HBr \\ R\cdot + Br_2 \longrightarrow RBr + Br\cdot \end{array} \right\} \text{propagation}$$

$$\left. \begin{array}{l} R\cdot + Br\cdot \longrightarrow RBr \\ R\cdot + R\cdot \longrightarrow R\text{–}R \end{array} \right\} \text{termination}$$

Any determination of the quantum yield of a process will depend on an accurate determination of the number of molecules reacting (or formed) and the number of photons absorbed. While the first quantity can be obtained by chemical analysis, a measure of the number of photons absorbed is best achieved by comparison with a reaction system whose quantum yield has been accurately determined over a wide wavelength range. This is achieved using thermopile–galvanometer techniques and such a reaction system is known as an *actinometer*.[7] Actinometers are available for the determination of quantum yields in the vapour phase and in solution. The most widely used liquid-phase actinometer consists of a sulphuric acid solution of potassium trisoxalatoferrate. On irradiation over a wide wavelength range (c. 250–500 nm), iron(III) is reduced to iron(II) while the oxalate undergoes simultaneous oxidation. Neither the iron(II) nor its oxalate complex absorb radiation in the operational wavelength range. The amount of iron(II) produced is determined spectrophotometrically after conversion into the 1,10-phenanthroline iron(II) complex which has a high absorption extinction coefficient. The number of photons absorbed by a given solution-phase reaction is determined in the following way. Firstly, the concentration of $K_3Fe(C_2O_4)_3$ in the actinometer is so adjusted that the solution absorbs exactly the same amount of radiation as the reaction solution whose quantum yield is required. This is achieved by the use of light tubes. These solutions are then irradiated under identical conditions and the quantity of iron(II) produced is determined. The number of photons absorbed by the actinometer can then be calculated from an accurate knowledge of the quantum yield for iron(III) to iron(II) conversion at the wavelength used and the number of iron(II) ions produced. By virtue of the conditions employed, the number of photons absorbed by the reaction system whose quantum yield is required must be the same.

1.4   The transfer of excitation energy – sensitisation and quenching[7,8]

Excitation of a ground state molecule by energy transfer from another excited species is termed *sensitisation*, and the deactivation of the excited species is termed *quenching*. Both sensitisation and quenching play an important role in organic photochemistry. The ease and efficiency of ISC varies from compound to compound and for many species, particularly alkenes, a significant triplet state population cannot be obtained in this way. The triplet state, however, can be obtained by the transfer of excitation energy from a different molecule in the triplet state. In any transfer of electronic energy between an excited species and a ground state species, the overall spin angular momentum does not change. For an excited state donor molecule D (sensitiser) and a ground state acceptor molecule A (quencher), three modes of energy transfer, (a)–(c), are possible (the arrows in parentheses represent the electron spins):

$$D_{S_0}(\uparrow\downarrow) \xrightarrow{h\nu} D_{S_1}(\uparrow\downarrow) \xrightarrow{\text{ISC}} D_{T_1}(\uparrow\uparrow)$$

(a) $D_{S_1}(\uparrow\downarrow) + A_{S_0}(\uparrow\downarrow) \longrightarrow D_{S_0}(\uparrow\downarrow) + A_{S_1}(\uparrow\downarrow)$

(b) $D_{T_1}(\uparrow\uparrow) + A_{S_0}(\uparrow\downarrow) \longrightarrow D_{S_0}(\uparrow\downarrow) + A_{T_1}(\uparrow\uparrow)$

(c) $D_{T_1}(\uparrow\uparrow) + A_{T_0}(\downarrow\downarrow) \longrightarrow D_{S_0}(\uparrow\downarrow) + A_{S_1}(\uparrow\downarrow)$

Scheme 1.1

Initial excitation of the sensitiser (D) produces the first excited singlet state (D_{S_1}), and quenching of this by a singlet ground state acceptor (A_{S_0}) produces ground state donor (D_{S_0}) and singlet excited acceptor (A_{S_1}) (route (*a*) scheme 1.1). These processes occur with conservation of electron spin. With triplet excited donor and singlet ground state acceptor, sensitisation of the acceptor to its triplet state can occur (route (*b*) scheme 1.1). When a triplet ground state species, such as oxygen, is present, quenching of the excited donor triplet state to singlet ground state occurs with formation of acceptor (oxygen) singlet state (route (*c*) scheme 1.1).

Two types of excitation energy transfer are known, namely *resonance-excitation* transfer and *exchange-energy* transfer. The former can occur when the absorption spectrum of the acceptor overlaps the emission spectrum of the donor. Energy transfer is known to occur with the donor and acceptor separated by as much as 5 nm. This type of process is most favoured when deactivation of the donor molecule and excitation of the acceptor molecule are independently spin-allowed, as shown in route (*a*) of scheme 1.1. Singlet sensitisation can therefore be effected by resonance-excitation transfer. Exchange-energy transfer requires a molecular collision or near-collision such that the electron clouds of the donor and acceptor overlap. In the region of overlap the electrons are indistinguishable and the excited electron on the donor could itself be transferred to the acceptor molecule. Exchange-energy transfer requires the conservation of spin angular momentum and hence the processes (*a*)–(*c*) are allowed.

The transfer of excitation energy by molecular collisions provides the best method for obtaining a triplet state population for molecules where ISC is inefficient. This is illustrated by route (*b*) scheme 1.1, and in fig. 1.8,

Fig. 1.8. Collisional energy transfer – triplet sensitisation. D = donor; A = acceptor.

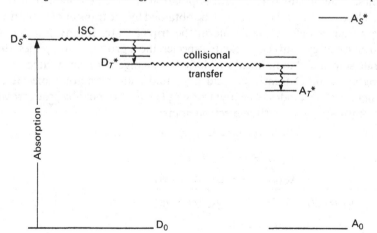

Table 1.1. *Benzophenone sensitisation of naphthalene*

	λ_{max} (nm)	E_{S_1} (kJ mol^{-1})	E_{T_1} (kJ mol^{-1})	$\tau_{T_1}^{\dagger}$ (s)	$\Phi_{ISC}^{\dagger\dagger}$
Naphthalene	320	385	255	10^{-4}	0.7
Benzophenone	367	314	289	10^{-4}	1.0

† Triplet state half-life in seconds.
†† Quantum yield for ISC.

where the acceptor is triplet sensitised. The triplet energy of the donor molecule must be greater than that of the acceptor otherwise sensitisation would not take place. In addition the donor triplet must be sufficiently long lived to enable collision to occur (a lifetime of $c.\ 10^{-6}$ s is adequate if the concentration of acceptor is 0.01 mol dm^{-3}) and must be photochemically stable. It is desirable that the first excited singlet of the acceptor be higher in energy than that of the donor otherwise singlet excitation ($A_0 \rightarrow A_S^*$) and singlet energy transfer ($D_S^* \rightarrow A_S^*$) could compete. Since the donor triplet level is higher in energy than that of the acceptor molecule, triplet sensitisation will produce the acceptor triplet state in a high vibrational level and vibrational cascade will follow. By careful choice of the donor molecules, exchange-energy transfer can provide a means of selectively populating the excited triplet state of the acceptor molecule.

One of the most frequently used triplet sensitisers is benzophenone which has a triplet energy of 289 kJ mol^{-1}, but a whole range of sensitisers of known triplet energies is available.[7,8] Benzophenone can be used to effect population of the triplet state of naphthalene and the data for the excited singlet and triplet states of naphthalene and benzophenone are given in table 1.1 and demonstrate conditions satisfactory for sensitisation. Since the energy and geometry of the excited singlet and triplet states for a given molecule differ, the photochemical reactions of molecules from these states are frequently entirely different.

The process of sensitisation allows the population of an excited state otherwise difficult to attain, while the reverse process, the quenching of an excited species by a ground state molecule, can remove an excited species before reaction can occur. The most common quencher is molecular oxygen, a ground state triplet. Collision of oxygen with a triplet state species can result in quenching of the triplet and formation of an excited singlet (route (c) scheme 1.1). The energy required to excite oxygen to its

$$D(\uparrow\uparrow) + O_2(\downarrow\downarrow) \longrightarrow D(\uparrow\downarrow) + O_2(\uparrow\downarrow)$$

lowest excited singlet state is $c.\ 92$ kJ mol^{-1}, a value less than the triplet energy of most organic molecules (for benzene $E_T = 351$ and for anthracene $E_T = 196 \, \text{kJ mol}^{-1}$ respectively). This low value is a consequence of the lowest lying excited singlet state of oxygen being obtained by spin inversion and electron pairing in the degenerate π^* levels, and is an efficient process.

$$O_2(\uparrow \ \uparrow)_{\pi^*} \longrightarrow O_2(\uparrow\!\!\downarrow \ -)_{\pi^*}$$

Ground state oxygen has been shown to quench triplet acetone at almost every molecular collision. The quenching of an excited singlet species by oxygen is also efficient. While the spin conservation rule predicts quenching of the donor singlet to its triplet state, this is rarely observed and the donor molecule is normally deactivated to the singlet ground state.

$$D_{S_1}(\uparrow\!\!\downarrow) + O_2(\uparrow\uparrow) \longrightarrow D_{T_1}(\uparrow\uparrow) + O_2(\uparrow\!\!\downarrow)$$

It has been suggested that this is a result of an increase in spin–orbit coupling by paramagnetic oxygen with a consequent increase in the number of spin-forbidden deactivations. The involvement of triplet oxygen in a charge-transfer complex with the excited singlet species has also been suggested to account for the spin-forbidden deactivation. However, singlet quenching is much less important than triplet quenching since the equilibrium concentration of an excited singlet species is generally much lower than that of the corresponding triplet as a result of the different lifetimes of the excited states of oxygen. Frequently the singlet oxygen generated by quenching of an excited triplet state diverts the course of a reaction. For example, in the absence of oxygen, anthracene photodimerises while in the presence of oxygen an endoperoxide is formed. The latter product arises by triplet quenching with subsequent addition of singlet oxygen to ground state anthracene in a Diels–Alder process.

If a photochemical reaction is thought to proceed from the triplet state, the introduction of a suitable triplet quencher should prevent, retard or divert the reaction. Among the quenchers used for this purpose are dienes, which usually have a high singlet and low triplet energy. For example, the dimerisation of cyclopentenone is inhibited by penta-1,3-diene (E_T 247 kJ mol^{-1}). Reactions which do not proceed except in the presence of a triplet sensitiser proceed from the triplet state. Direct evidence for the

presence of a triplet state species in a reaction can sometimes be obtained from an e.p.r. spectrum determined during the course of the reaction. If a reaction is insensitive to triplet quenching and forms different products on sensitised irradiation, it can be inferred that the initial reaction involves a singlet excited state.

In many photochemical reactions the excited species decompose to free radicals, the presence of which can be determined by free radical traps such as alkenes, nitric oxide and hydrocarbons.

References

1. Gillam, A. E. & Stern, E. S., *Introduction to Electronic Absorption Spectroscopy in Organic Chemistry*, 3rd ed, eds., Stern, E. S. & Timmons, C. J., Arnold, 1970.
2. Murrell, J. N., *The Theory of Electronic Spectra of Organic Molecules*, Chapman and Hall, 1971.
3. Jaffé, H. H. & Orchin, M., *Theory and Applications of Ultraviolet Spectroscopy*, Wiley, 1962.
4. Becker, R. S., *Theory and Interpretation of Fluorescence and Phosphorescence*, Wiley, 1969.
5. Alger, R. S., *Electron Paramagnetic Resonance*, Wiley, 1968.
6. Scaiano, J. C., *Acc. Chem. Res.*, 1982, **15**, 252.
7. Turro, N. J., *Modern Molecular Photochemistry*, Benjamin/Cummings, 1978.
8. Calvert, J. G. & Pitts, J. N., *Photochemistry*, Wiley, 1966.

2   Intramolecular reactions of the alkene bond

The photochemistry of alkenes has attracted a great deal of attention and is now relatively well understood. In this chapter the general types of intramolecular processes involving alkenes are covered while dimerisations and intermolecular addition reactions are discussed separately (see chapter 4). The energy of the triplet state T_1 of an alkene is lower than that of the excited singlet state S_1. However, intersystem crossing (ISC) is generally inefficient and in order to examine reactions of the triplet excited state, population of this state must be achieved by sensitisation. The photolysis products from an intramolecular reaction of an alkene are dependent on the nature of the alkene, the number of sites of unsaturation in the molecule and their structural relationship to one another, and frequently on the excited state from which the reaction proceeds. While the intramolecular reactions of alkenes fall under the general heading of 'rearrangements' these rearrangement reactions can be subdivided into geometrical isomerisation, electrocyclic reactions, sigmatropic reactions and intramolecular cycloadditions.

2.1   Geometrical isomerisation[1-7]

The phenomenon of *cis–trans* or *(E)–(Z)* isomerisation is well established not only for thermal ground state reactions, but also for excited state processes.[1-7] Furthermore, the reaction appears to be general for all types of double-bond linkages capable of exhibiting geometrical isomerisation and, when effected with light, will lead to a photostationary state.

The ease with which alkenes undergo geometrical isomerisation on photolysis has been known for a considerable time, but the details of the reaction continue to attract attention. In a ground state, or thermally induced isomerisation the reaction is considered to proceed by way of a non-planar transition state common to both *cis*- and *trans*-isomers. This transition state collapses to give a greater proportion of the thermodynamically more stable isomer, usually the *trans*-isomer. A similar situation can apply in the photochemical reaction where population of the S_1 and T_1 states is followed by vibrational cascade which is associated with a twisting of the molecule about the carbon–carbon σ-bond of the alkene.[1–3] The energy *minimum* of both the S_1 and T_1 excited states corresponds to a structure in which a rotation of $90°$ has occurred and where the adjacent p-orbitals are orthogonal. This structure is frequently referred to as the '*p*' *state* and corresponds in geometry to the energy *maximum* of the ground state, S_0, of the alkene. Radiationless transition by internal conversion (IC) (from S_1) or ISC (from T_1) from the '*p*' state will deliver an alkene ground state high in internal energy which will collapse to the *cis*- and *trans*-isomers (cf. path (*a*), scheme 2.1). In this way an equilibrium between the

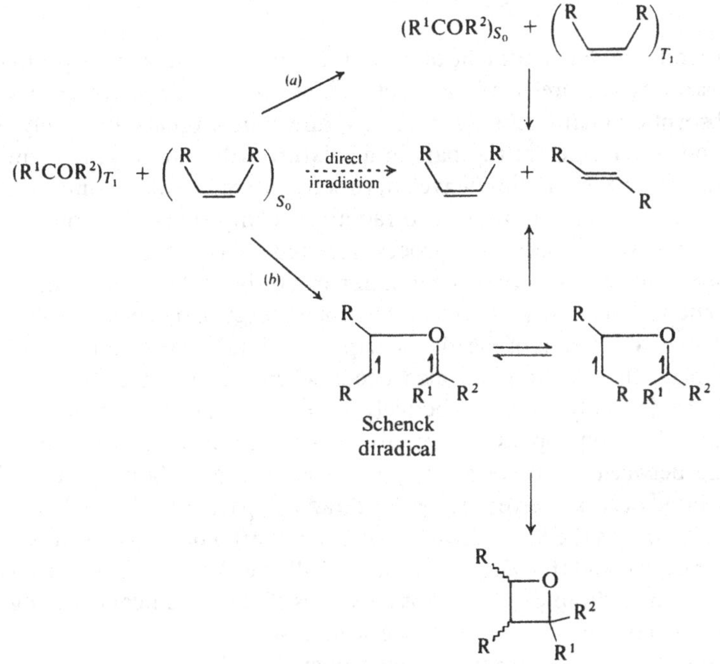

Scheme 2.1

geometrical isomers, termed the *photostationary state*, is established. Irradiation of a pure *cis*- or *trans*-isomer, or a *cis–trans* mixture will lead to the *same* photostationary state of alkene isomers. Because of the difficulties associated with direct irradiation below 200 nm the number of studies of simple alkenes is not yet great. However, irradiation of *trans*-1,2-dideuterioethene in the vapour phase at 147 and 180 nm affords the *cis*-

Table 2.1. *Sensitiser energy and the pent-2-ene photo-stationary composition*

Sensitiser	E_T (kJ mol^{-1})	% cis	% trans
PhCOPh	285	15	85
PhCOCH$_3$	310	16	84
CH$_3$COCH$_3$	335	40	60
C$_6$H$_6$	350	50	50

isomer while similar treatment of *cis*-but-2-ene delivers the *trans*-product, demonstrating the preferred mode of collapse of the respective 'p' states. The absorption characteristics of the *cis*- and *trans*-isomers are rarely the same and one isomer of the pair in a mixture will undergo preferential excitation. If the irradiation wavelength were filtered to allow only one of the isomers in a mixture to absorb radiation, complete conversion to the other isomer would occur – a process referred to as *optical pumping*.

Photosensitised *cis–trans* isomerisation can be brought about from T_1 of the alkene as described above when the triplet energy of the sensitiser allows for effective population of the alkene triplet (path (*a*), scheme 2.1). Perhaps surprisingly, the reaction can also occur when the triplet energy of the sensitiser is formally too low effectively to populate the alkene triplet. This abnormal situation applies for a number of ketone sensitisers. In such cases bonding between the carbonyl oxygen atom and a carbon atom of the double bond occurs to give a triplet 1,4-diradical (path (*b*), scheme 2.1). This species, known as the *Schenck diradical*, can undergo bond rotation before cleavage occurs and thus deliver isomerised alkene. An alternative pathway for the Schenck diradical to collapse involves ISC and radical coupling to give the corresponding oxetanes (see section 4.3).

Because there is more than one mechanism for triplet sensitised isomerisation, the composition of the photostationary state that is attained in any instance is dependent upon the triplet energies of the sensitiser and acceptor alkene. This is illustrated for the pent-2-enes (table 2.1) for which E_T is *c*. 330 kJ mol^{-1}. The excited T_1 state of benzene can populate the alkene triplet and give a 1:1 mixture of isomers whereas benzophenone and acetophenone, with lower T_1 energies, give products, via the Schenck 1,4-diradical, which not surprisingly collapses preferentially to the *trans*-isomer. In fact these latter sensitisers afford a photostationary state of an alkene the composition of which approaches that of the thermodynamic equilibrium. Not surprisingly, the more conjugated the alkene the easier it is to effect singlet state photoisomerisation. Many conjugated systems have been examined but none more so than the *cis–trans* isomerisation of stilbenes.[1,2] The steric interference between the *ortho*-hydrogens of the aromatic rings of *cis*-stilbene results in some rotation of the phenyl rings out of plane with the remainder of the molecule and there is consequently a less effective overlap of the π-system. This is reflected in the ultraviolet absorption characteristics of the isomers. The *trans*-isomer absorbs light of wavelength 294 nm ($\varepsilon = 2400 \, \mathrm{m^2 \, mol^{-1}}$) while the *cis*-isomer absorbs at 278 nm ($\varepsilon = 935 \, \mathrm{m^2 \, mol^{-1}}$).

cis-stilbene   *trans*-stilbene

Direct irradiation of *trans*-stilbene gives a transoid singlet excited state from which ISC is inefficient. Rotation in this excited species gives the twisted 'p' state but, for *trans*-stilbene, this rotation is endothermic (*c.* 17 kJ mol^{-1}). Whether the 'p' state is a singly or doubly excited species is not known, but collapse from this 'p' state by conventional deactivation processes gives almost equal amounts of the *cis*- and *trans*-stilbene. In like manner direct absorption by *cis*-stilbene results in the 'p' state, but in this case the rotation from S_1 is exothermic because steric interference is minimised by the rotation. Substitution at the *ortho*-positions of the aromatic rings enhances steric interference in the *cis*-ground state and results in even more facile twisting from the S_1 excited state to the 'p' state. By incorporation of heavy atom substituents, e.g. bromine, or groups which can conjugate with the double bond and lower the energy of the n–π^* excited state, e.g. COR, ISC is enhanced and decay to the ground state geometrical isomers occurs for these substrates from a triplet 'p' state.

If, in a sensitised process, the donor species fulfils the conditions for effective sensitisation, and its triplet energy is greater than that of either of the geometrical isomers, then triplet energy transfer to both the *cis*- and *trans*-isomers will occur. The initially formed *cis*- and *trans*-triplet excited species undergo distortion to the common 'p'-triplet which, on collapsing to the ground state, affords a mixture of isomers. Because the sensitiser can excite both isomers, the proportion of *cis*-isomer in the photostationary state from a sensitised reaction is lower than that obtained from direct irradiation since on direct irradiation the isomer with the higher extinction coefficient, namely the *trans*-stilbene, undergoes preferential excitation. The *cis–trans* isomerisation of 1,2-diphenylpropene has been studied for a series of sensitisers of known triplet energies. Sensitisers of high energy give photostationary states with approximately the same composition of the geometrical isomers (55% *cis*), while direct photolysis results in a higher proportion of *cis*-isomer (72%). As the sensitiser energy is reduced anomalous results are observed which have been accounted for by 'non-classical' energy transfer in violation of the Franck–Condon principle, and direct production of the 'p'-triplet state.

Many other examples of geometrical isomerisation are known which are

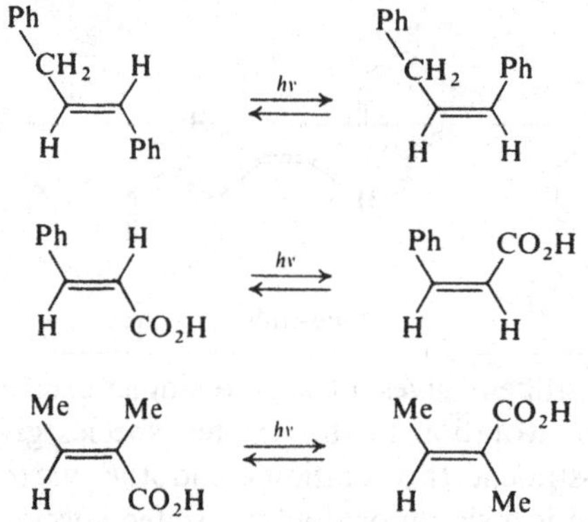

further exemplified by the azo (1) and oxime (2) derivatives shown below. Either direct or sensitised irradiation of the *trans*-azo compounds (1) leads to the less stable *cis*-isomers which are frequently thermally labile and

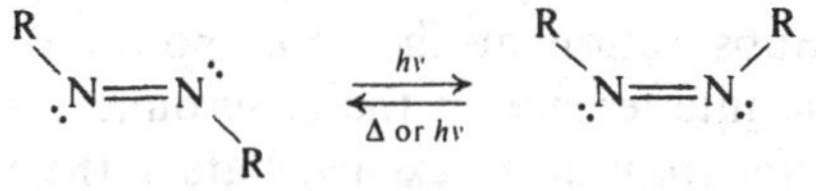

trans-(1)   *cis*-(1)

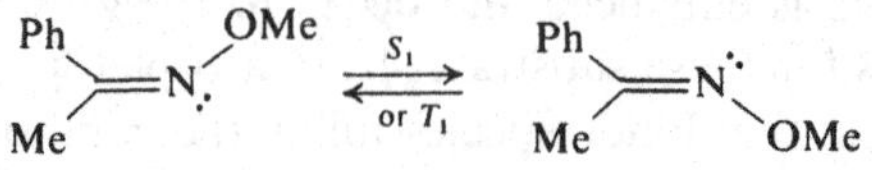

syn-(2)   *anti*-(2)

either revert to *trans*-(1) or undergo fragmentation into radicals (section 5.6). The oxime ethers (2) separately lead on direct or sensitised photolysis to a photostationary state containing *c.* 66% of the *syn*-isomer.

One of the particularly valuable aspects of *cis–trans* photoisomerisation is that it provides an entry into *trans*-cycloalkenes. An eight-membered ring is the minimum size to accommodate a *trans*-double bond as a stable entity at ambient temperature and *trans*-cyclooct-2-enone is obtained from n–π* excitation (> 300 nm) of the *cis*-isomer. An analogous reaction occurs for *cis*-cyclohept-2-enone but in this instance the strain energy of the *trans*-product precludes its isolation except at low temperatures. Triplet energy transfer reactions of 1-phenyl cyclooctene, cycloheptene and cyclohexene likewise lead to the formation of *trans*-isomers. The cyclohexene derivative (3), whose existence has been confirmed by flash photolysis kinetic

(3)

measurement techniques, is the smallest ring-system to sustain a *trans*-double bond as a short-lived intermediate but this suffers the addition of methanol under the conditions employed.[4,5]

The solution-phase photolysis of dienes and trienes either directly or under sensitised conditions results in geometrical isomerisation. However, the singlet state reaction is frequently complicated by competing electrocyclic, sigmatropic or intramolecular cycloaddition reactions. These processes often give products which do not absorb radiation at the wavelength employed and hence are stable to the reaction conditions; the reaction is thus driven to completion. For example, direct absorption by *cis*- or *trans*-penta-1,3-diene results in *cis–trans* isomerisation but the *trans*-isomer competitively gives 3-methylcyclobutene by an electrocyclic reaction (see section 2.2). By comparison the triplet state reactions of conjugated dienes and trienes frequently provide for uncomplicated geometrical isomerisation. Taken with the results of direct excitation, this shows that ISC in these compounds is inefficient, and, in fact, dienes can be used as triplet quenchers. A further point of interest is that when geometrical isomerisation is possible at both of the double bonds of a diene the singlet state reaction causes rotation about only one of the double bonds. The analogous triplet state process involves rotation about both alkene bonds as illustrated for hexa-2,4-diene; radical delocalisation is

cis,trans	*trans,trans*	*cis,cis*	*cis,trans*

possible in the longer-lived T_1 state allowing time for rotation about both of the alkene double bonds.

The photochemistry of vision is triggered by absorption of a photon and *cis–trans* isomerisation.[6] The conjugated polyenal, 11-*cis*-retinal and the protein opsin combine in the retina to give the red-purple 11-*cis*-imine, rhodopsin (scheme 2.2). When absorption of light occurs the *cis*-double bond is isomerised to give the yellow all *trans*-metarhodopsin II and a nerve impulse is triggered. Unlike its 11-*cis*-imine isomer the all *trans* product does not fit the site on the protein surface and the carbon–nitrogen double bond is exposed and hydrolysed to give all *trans*-retinal and opsin. The all *trans*-retinal can be converted back to the 11-*cis*-isomer in the retina by an enzyme and light of much shorter wavelength. In bright light metarhodopsin II can be converted to the 11-*cis*-imine, rhodopsin, by *trans–cis* isomerisation of the 11,12-carbon–carbon double bond. This is an example of a reversible photoinduced isomerisation where colour changes occur upon *cis–trans* isomerism. Such systems are termed *photochromic* systems. Photochromism can be brought about by a variety of photochemical reaction types (e.g. photoenolisation, section 3.1). Nevertheless, *cis–trans* isomerisation accounts for the photochromic behaviour or colour changes of a large number of examples, including dyes structurally related to the indigo ring skeleton.[7] Photochromism of the thioindigo dyes (4; Y = S) with a variety of aromatic substituents have been well studied and colour changes are also observed for the derivatives of (4) with Y = NR, O, or Se.

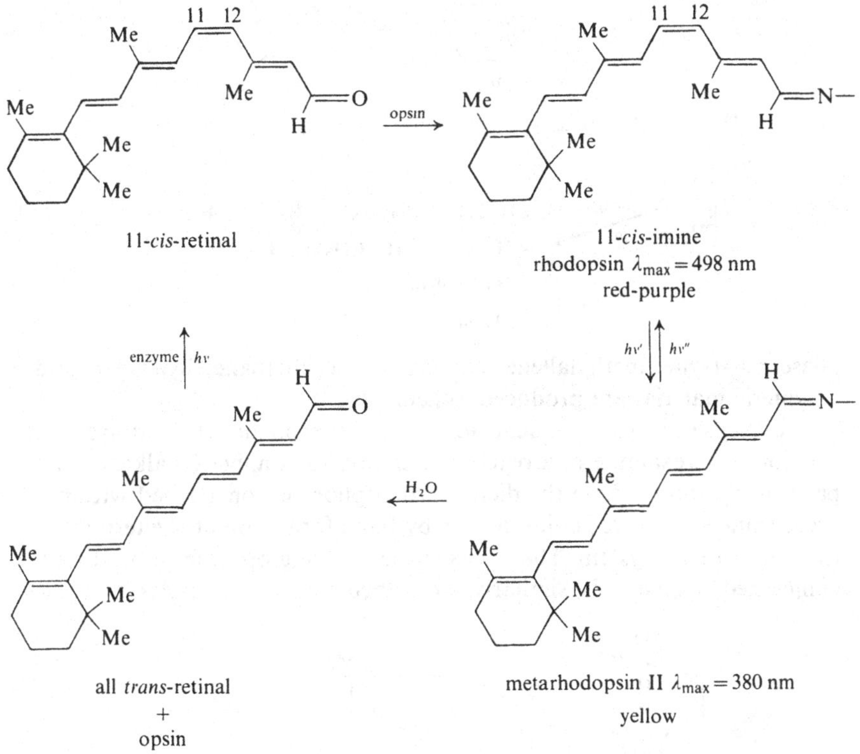
11 12
Me
Me
Me
=O
H
Me
Me
opsin
11-cis-retinal
11-cis-imine
rhodopsin $\lambda_{max}=498$ nm
red-purple
enzyme hν
hν′ hν″
H_2O
all trans-retinal
+
opsin
metarhodopsin II $\lambda_{max}=380$ nm
yellow
=N—

Scheme 2.2

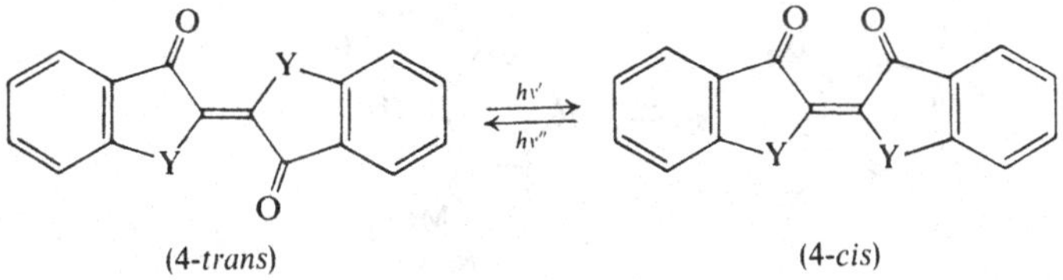
O
Y
hν′
hν″

(4-*trans*) (4-*cis*)

2.2 Cyclisation reactions of conjugated alkenes[8–18]

The photochemical reactions of conjugated alkenes depend to a large extent on the excited state populated, and on the phase in which the reaction is performed. Singlet excitation generally leads to intramolecular processes whereas dimerisation and intermolecular addition reactions are more common from the triplet excited state. Reaction in the vapour phase at low pressure often leads to greater fragmentation than in solution because the high vibrational levels populated on excitation are not as rapidly deactivated by collisional processes. The solution-phase photolysis of butadiene leads to cyclobutene and bicyclobutane, whereas in the vapour

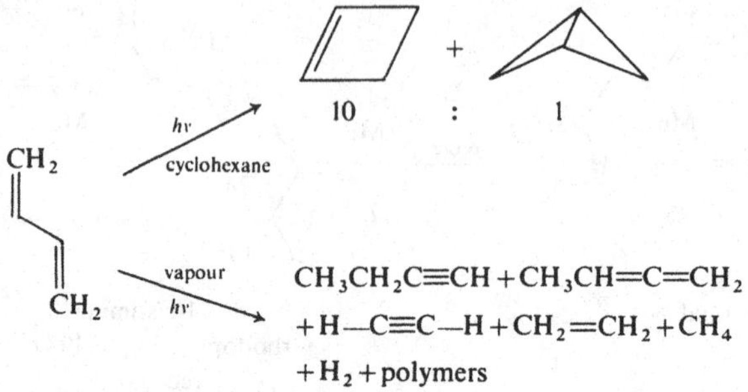

$$CH_3CH_2C\equiv CH + CH_3CH=C=CH_2$$
$$+ H-C\equiv C-H + CH_2=CH_2 + CH_4$$
$$+ H_2 + \text{polymers}$$

Scheme 2.3

phase but-1-yne, methylallene, ethyne, ethene, methane, hydrogen and polymeric material are produced (scheme 2.3).

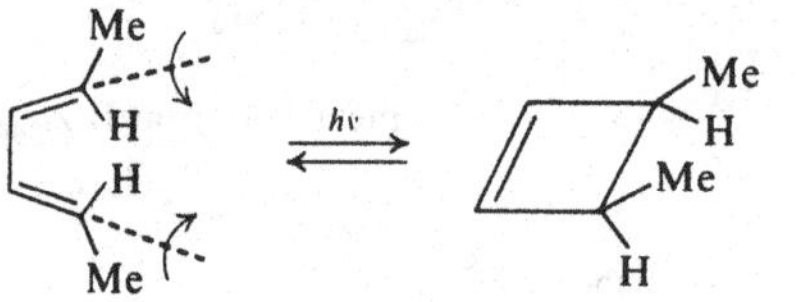

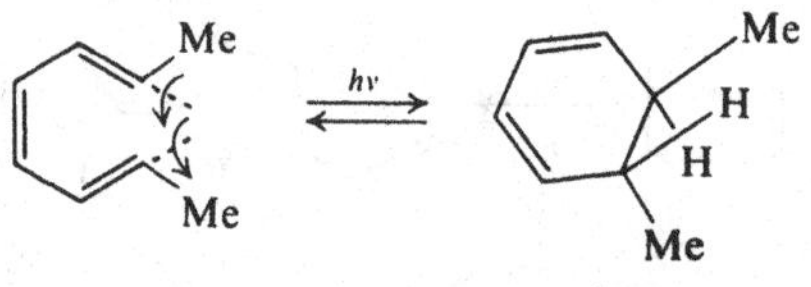

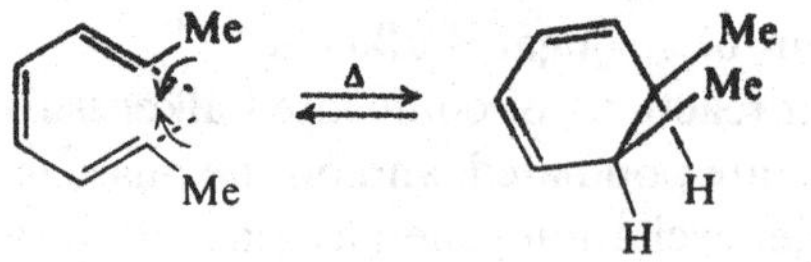

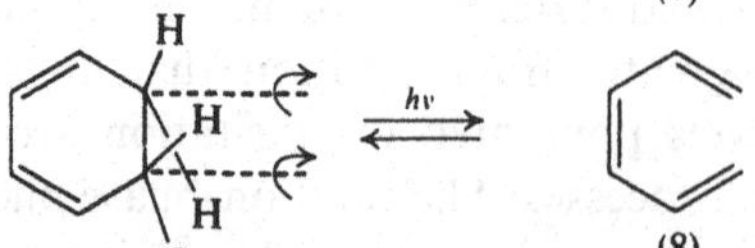

The cyclisation of butadiene to cyclobutene on direct photolysis in solution is an example of a reaction common to conjugated alkenes and particularly those where the diene chromophore is constrained within a cyclic framework. Cyclisation occurs by bond formation at the termini of the conjugated π-system. The reverse process – ring opening to produce a conjugated alkene – is similarly well documented. Examples of these reactions are numerous (see, for example, (5)–(8) opposite), with the position of equilibrium between 'ring-opened' and 'ring-closed' compounds depending on the particular structure of the substrate. A characteristic of all these reactions is the remarkable stereospecificity of ring opening and closure, only *cis*-3,4-dimethylcyclobutene (5) being formed from *trans,trans*-hexa-2,4-diene, and only *trans*-5,6-dimethylcyclohexa-1,3-diene (6) from *trans, cis,trans*-octa-2,4,6-triene. When cyclisation of this latter substrate is effected thermally, *cis*-5,6-dimethylcyclohexa-1,3-diene (7) is produced in contrast to the photochemically induced cyclisation in which the *trans*-dimethyl isomer (6) is obtained. The rationalisation by Woodward and Hoffmann of the need for *conservation of orbital symmetry in a concerted reaction* accounts for the stereospecificity of these cyclisations.[8–10]

The ring opening and closure reactions of conjugated alkenes are termed *electrocyclic* reactions. Such a reaction is defined as '*the formation of a single bond between the termini of a conjugated polyene, or the converse process*'. The orbital symmetry arguments of Woodward and Hoffmann are a consequence of the application of molecular orbital theory to reacting systems, with the basic premise that *in a concerted reaction orbital symmetry is conserved*. The mode of closure and the stereochemical outcome of the butadiene–cyclobutene interconversion is determined by the molecular orbitals most involved in the reaction. Those most involved in the butadiene–cyclobutene transformation are shown in fig. 2.1 where the molecular orbitals are represented by overlapping atomic orbitals. The butadiene π-orbital with the lowest number of nodes† is the lowest energy π-orbital, and as the number of nodes increases, the energy correspondingly increases. The electrons are fed into the orbitals in the usual way; thus in the ground state butadiene has the electron configuration $\psi_1{}^2\psi_2{}^2$ and ψ_2 is the *highest occupied molecular orbital* (HOMO). Excitation involves the promotion of an electron from the ground state HOMO, (ψ_2), to the ground state *lowest unoccupied molecular orbital* (LUMO) (ψ_3) and thus the first excited state of butadiene has the configuration $\psi_1{}^2\psi_2{}^1\psi_3{}^1$, *viz* the ground state LUMO is occupied in the first excited state and becomes the HOMO for the first excited state.

A molecular orbital analysis of the cyclisation of butadiene to cyclo-butene must include all four molecular orbitals of butadiene (ψ_1–ψ_4) and the σ-, π-, π*- and σ*-orbitals of cyclobutene (fig. 2.1) since these are the orbitals most involved in the reaction. This is not to say that other lower energy σ-orbitals are not important to the energy requirement of cyclisation since these bonds are also involved with bond angle deformations in the cyclisation. In bringing about the electrocyclic ring closure of butadiene it is necessary for twisting or rotation to occur about the C_1–C_2 and the C_3–C_4

---

† A node is defined as a point in space where there is zero probability of finding an electron.

bonds. When the rotation occurs in the same direction the process is termed *conrotation* and when the rotation about these bonds is in opposite directions the process is termed *disrotation*. The cleavage of cyclobutene to give butadiene can likewise occur by a conrotatory or disrotatory process.

Fig. 2.1. Molecular orbitals involved in the butadiene-cyclobutene transformation.

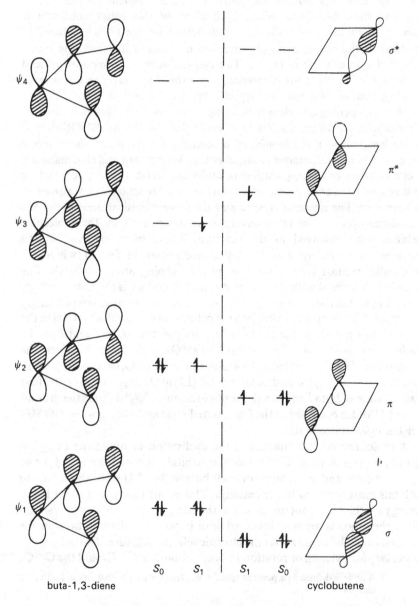

buta-1,3-diene   cyclobutene

The two modes of conrotation

The two modes of disrotation

As the diagrams show there are two modes of conrotation and two modes of disrotation possible for every electrocyclic reaction.

Conrotatory cyclisation of butadiene preserves a two-fold axis of symmetry throughout the reaction (fig. 2.2) while disrotation preserves a plane of symmetry (fig. 2.3). All of the orbitals involved in the thermal and photochemical reactions can be classified as either antisymmetric (A) or symmetric (S) with respect to these symmetry elements. For example, the ψ_1-molecular orbital of butadiene is symmetric (S) with respect to the plane of symmetry, and antisymmetric (A) with respect to the two-fold axis of symmetry. A correlation diagram for conrotatory or disrotatory cyclisation of butadiene is obtained by joining orbitals of like symmetry. The *quantum mechanical non-crossing rule*, which states that *only levels of unlike symmetry are allowed to cross*, must be observed in constructing the diagram. The diagram so constructed shows that for conrotatory closure the ground state bonding levels of butadiene (ψ_1, ψ_2) and cyclobutene (σ, π) correlate directly (fig. 2.2). Thermal cyclisation of butadiene to cyclobutene and thermal cleavage of cyclobutene to butadiene are therefore predicted to proceed by conrotation. For disrotatory closure it is the first excited state of butadiene ($\psi_1{}^2, \psi_2{}^1, \psi_3{}^1$) that correlates directly with the first excited state of cyclobutene ($\sigma^2, \pi^1, \pi^{*1}$) (fig. 2.3). Photocyclisation of butadiene to cyclobutene, and photocleavage of cyclobutene to butadiene, are therefore predicted to proceed by disrotation.

The correlation diagrams show the transformation of the various molecular orbitals of starting material into those of product. From the correlation diagram for the conrotatory interconversion 1,3-butadiene–cyclobutene (fig. 2.2) the ψ_2-molecular orbital is transformed into the σ-orbital of cyclobutene and vice versa. The phases of the ψ_2-orbital are such that conrotation (in either sense) results in a bonding interaction as the terminal lobes of the orbital interact in forming product. Because the σ-bond is formed from the HOMO, the phases of the HOMO can be regarded as dictating the stereochemical course of rotation about C_1–C_2 and C_3–

Fig. 2.2. Correlation diagram for the conrotatory interconversion of buta-1,3-diene–cyclobutene – thermally favoured.

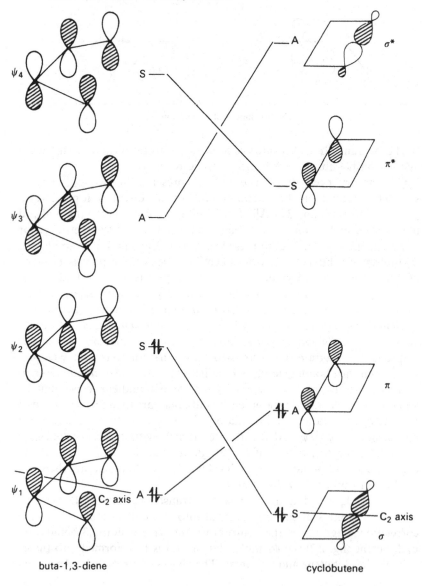

buta-1,3-diene   cyclobutene

C_4 during the reaction. The ψ_1-molecular orbital evolves into the π-orbital of the cyclobutene and it is notable that the phases of the ψ_1-orbital at C_2 and C_3 are appropriate to form the π-orbital of cyclobutene.

For the disrotatory interconversion of butadiene and cyclobutene, which

Fig. 2.3. Correlation diagram for the disrotatory interconversion of buta-1,3-diene–cyclobutene – photochemically favoured.

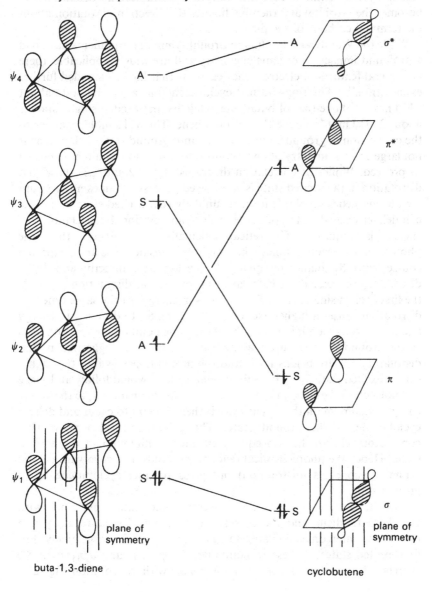

is the allowed photochemical pathway, the σ-bond in cyclobutene evolves from the ψ_1-moleuclar orbital (fig. 2.3); the ψ_2-molecular orbital develops into the π^*-orbital and the ψ_3-orbital into the π-orbital. It is sometimes said that the stereochemical outcome of the reaction is governed by the terminal phases of the HOMO (ψ_3) but the correlation diagram demonstrates that this is coincidental since the ψ_1- and ψ_3-molecular orbitals have the same phases at the terminal carbons. It is the ψ_1-molecular orbital which becomes the σ-orbital and thereby dictates the direction of rotation about the terminal carbons of the diene.

While the Woodward–Hoffmann orbital symmetry arguments have had a profound impact on organic chemistry and are widely applicable, there are several features of electrocyclic reactions that have demanded further explanation.[11] The ring-strained cyclobutene has a ground state about 84 kJ mol^{-1} above that of butadiene, while its first excited singlet state is about 230 kJ mol^{-1} above that of butadiene. The 84 kJ mol^{-1} barrier to the conversion of ground state butadiene into ground state cyclobutene is not large and a symmetry-allowed conrotatory reaction would be expected to proceed. While the correlation diagrams (figs. 2.2 and 2.3) show that disrotation in the excited state is an allowed process this does not account for the energetics involved; it seems unlikely that excited state butadiene can deliver excited state cyclobutene since an additional 230 kJ mol^{-1} of energy is required. Theoretical calculations[11] suggest that the photochemical reaction takes what might be considered as an unexpected course with S_1 butadiene giving S_0 cyclobutene directly and by a disrotatory process. These calculations show that as disrotation proceeds the first excited singlet state, S_1, increases in energy. At the same time this disrotation causes a higher excited singlet state, S_x, to decrease in energy and pass through a minimum at about the same point on the reaction path as the ground state closure reaches a maximum. The suggestion is that disrotatory closure begins in S_1 and the molecule passes into the higher singlet excited state S_x by IC; vibrational cascade would follow and place the molecule at the energy minimum of S_x. Conversion from S_x to the (lower energy) maximum on the S_0 pathway is then thought to occur and deliver cyclobutene in the ground state. The calculations also show that conrotatory closure has no equivalent energy minimum in the excited states. Hence the photochemical reaction proceeds in accord with orbital symmetry concepts (disrotation) but gives product cyclobutene in its ground state.

Hexa-1,3,5-triene can undergo electrocyclic interconversion with cyclohexa-1,3-diene and the correlation diagrams for conrotatory and disrotatory interconversion are shown in figs. 2.4 and 2.5 respectively. The first excited states of these reactants ($\psi_1{}^2\psi_2{}^2\psi_3{}^1\psi_4{}^1$ and $\sigma^2\pi_1{}^2\pi_2{}^1\pi_2^{*1}$) correlate for photochemical interconversion with conrotation (fig. 2.4).

Fig. 2.4. Correlation diagram for the conrotatory interconversion of hexa-1,3,5-triene–cyclohexa-1,3-diene – photochemically favoured.

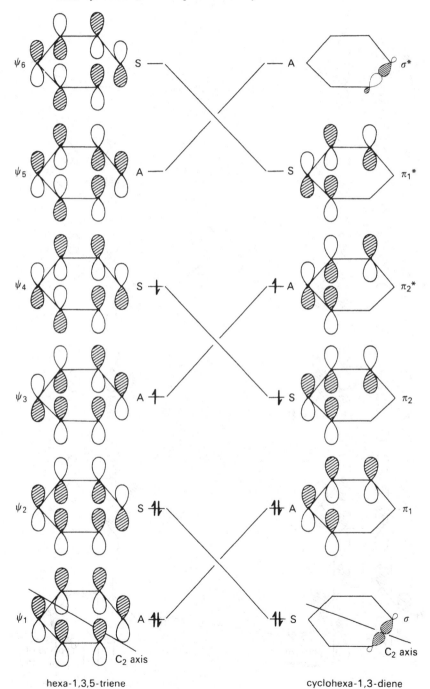

hexa-1,3,5-triene   cyclohexa-1,3-diene

34 *Intramolecular reactions of the alkene bond*

Fig. 2.5. Correlation diagram for the disrotatory interconversion of hexa-1,3,5-triene–cyclohexa-1,3-diene – thermally favoured.

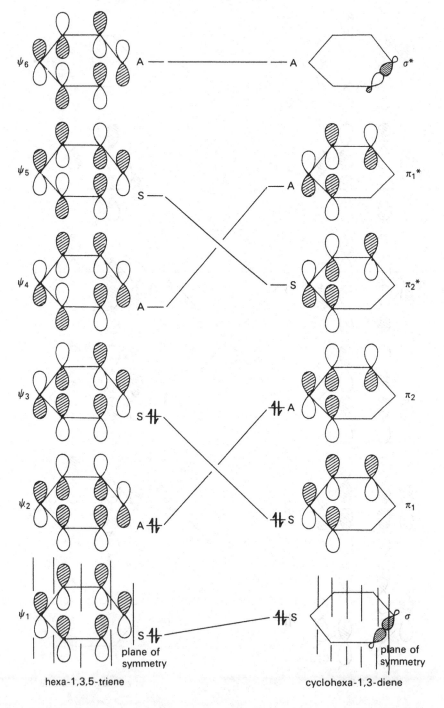

hexa-1,3,5-triene

cyclohexa-1,3-diene

Table 2.2. *Symmetry rules for electrocyclic reactions*

Number of electrons in conjugated ene	Thermal (ground state)	Photochemical (excited state)
$4n$	conrotatory	disrotatory
$4n+2$	disrotatory	conrotatory

Because cyclohexa-1,3-diene is not excessively strained the energy problems alluded to above do not arise in this photochemical interconversion. The ground state reactants ($\psi_1^2\psi_2^2\psi_3^2$ and $\sigma^2\pi_1^2\pi_2^2$) correlate for a disrotatory interconversion (fig. 2.5). The stereochemistry of each of the reactions appears to be controlled by the phases of the terminal orbitals of the HOMO. However, once again this is coincidental and results from the identity of the terminal phases of the HOMO with the phases of the orbital which transforms into the new σ-bond, or the converse. It should be noted that the mode of rotation required for hexatriene–cyclohexadiene conversion under either thermal or photochemical conditions is the *opposite* to that necessary for the butadiene–cyclobutene interconversion.

All conjugated alkenes or delocalised ions containing $4n$ π-electrons ($n =$ integer) provide correlation diagrams with the same outcome as butadiene. All systems with $(4n+2)$ π-electrons likewise behave as hexatriene. The converse process, ring opening to afford a conjugated alkene or a delocalised ion, will necessarily proceed with the same mode of rotation as dictated by the *law of microscopic reversibility*.[8,9] The mode of an electrocyclic reaction with the number of π-electrons involved in the process is summarised in table 2.2.

An alternative method of determining the stereochemistry of electrocyclic reactions is by the '*frontier orbital method*' which states that *the allowed mode in an electrocyclic reaction is one in which the HOMO of the reactant leads to a bonding orbital in the product*.[9-14] Inspection of the orbital correlation diagram for each of the electrocyclic reactions discussed (figs. 2.2–2.5) confirms that the HOMO leads to a bonding orbital in each of the allowed reactions. For example, in the butadiene–cyclobutene conversion the ψ_2-molecular orbital leads to a bonding σ-orbital for conrotation (fig. 2.2) while for disrotation the excited state HOMO, the ψ_3-orbital, leads to the bonding π-orbital (fig. 2.3).

The stereospecificity of an electrocyclic reaction can also be explained by considering the nature of the transition structure for the process.[12-14] In the transition state method one deals with atomic orbitals and the transition state structure is drawn as a series of overlapping atomic orbitals

Table 2.3. *Symmetry rules for electrocyclic reactions based on the transition state structure*

Orbital array	Number of electrons (n = integer)	Reaction Type	
		Thermal	Photochemical
Hückel	$4n + 2$	favourable	unfavourable
	$4n$	unfavourable	favourable
Möbius	$4n + 2$	unfavourable	favourable
	$4n$	favourable	unfavourable

(s and p) with phases inserted so as to minimise the number of phase changes in the participating orbitals. The number of phase inversions in the so constructed cyclic array are counted. If the cyclic array includes both lobes of a single p-orbital the necessary phase inversion across the two lobes is not counted. The transition state structure is classified as *Hückel* for *zero or an even number of phase inversions* or *Möbius* for an *odd number of phase inversions*. The number of electrons in the cyclic transition state structure are also counted. For a Hückel system if the number of electrons is $4n + 2$ (n = integer) the cyclic delocalisation is regarded as aromatic whilst a $4n$ electron system is regarded as antiaromatic. Thermally favoured reactions are those with aromatic ($4n + 2$) electron delocalisation and a Hückel orbital array; antiaromatic ($4n$) electron delocalisation, and a Möbius orbital array, also gives rise to a ground state reaction. For excited state reactions the converse applies and the outcome of the 'aromatic transition state structure' concept is shown in table 2.3. The conrotatory transition state structure for the butadiene–cyclobutene interconversion has a Möbius configuration with one phase inversion that counts and four electrons are involved in the transition. The reaction is allowed under thermal conditions. The transition state structure for disrotatory

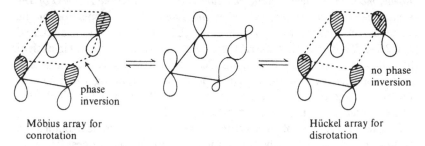

Möbius array for
conrotation

no phase
inversion

Hückel array for
disrotation

interconversion is a Hückel type with no phase inversions. Four electrons are again involved and therefore the transition state structure is unfavourable for a thermal, but favourable for a photochemical, reaction as is observed.

Woodward and Hoffmann have defined the allowed or non-allowed nature of electrocyclic reactions, and indeed pericyclic reactions in general, by a single rule. The reaction under consideration is considered in terms of suprafacial and antarafacial components. Reaction at the termini of a 2π- or 4π-system or indeed at the termini of a more extended π-system can be from the same face when the attack is termed *suprafacial*, or from opposite faces

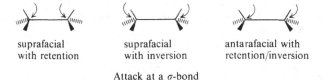

suprafacial attack at a π-system

antarafacial attack at a π-system

when the attack is referred to as *antarafacial*. This same terminology is applied to reactions at a σ-bond. If configuration is retained at each end of

suprafacial   suprafacial   antarafacial with
with retention   with inversion   retention/inversion

Attack at a σ-bond

the σ-bond the attack is *suprafacial*. If attack at each end of the bond results in inversion of configuration the attack is similarly *suprafacial*. However if attack at one end involves retention of configuration whilst that at the other end involves inversion then the attack is *antarafacial*. This is illustrated in fig. 2.6 for the conrotatory and disrotatory interconversions of butadiene–cyclobutene. The conrotatory cyclisation involves antarafacial addition of the 4π-electrons across the termini of the system; this is designated a $[_\pi 4_a]$ process. In similar manner the conrotatory ring opening requires addition to the opposite faces of the double bond $[_\pi 2_a]$, and suprafacial addition to the σ-bond $[_\sigma 2_s]$; this is termed a $[_\pi 2_a + _\sigma 2_s]$ process. The disrotatory closure and opening (fig. 2.6(*b*)) are $[_\pi 4_s]$ and $[_\pi 2_s + _\sigma 2_s]$ process respectively.

The *generalised Woodward–Hoffmann rule* states that *a ground state reaction is orbital symmetry allowed if the sum of the* $(4q+2)_s$ *and* $(4r)_a$ *components* (where q and r are integers) *is odd*. The $(4q+2)$ components are 2_s ($q=0$), 6_s ($q=1$), 10_s ($q=2$) and the $(4r)_a$ components 0_a ($r=0$), 4_a ($r=1$), 8_a ($r=2$) etc. For the conrotatory closure of fig. 2.6(a) there are *no* components of the $(4q+2)$, type and *one* component of the $(4r)_a$ type. The total number of components is *odd* and the reaction is thermally allowed. For the conrotatory ring cleavage only the $(_\sigma 2_s)$ component is of the $(4q+2)_s$ type and there is no $(4r)_a$ component; the reaction is thermally allowed. For the disrotatory closure of butadiene [fig. 2.6(b)] there are *no* components of the $(4q+2)_s$ or $(4r)_a$ types whilst for the disrotatory ring opening of cyclobutene *both* the $(_\pi 2_s)$ and $(_\sigma 2_s)$ components are of the $(4q+2)_s$ type. Thus there are *no* components for the forward reaction and an *even number* for the reverse reaction and the interconversion is *not* thermally allowed. The corollary to the generalised rule is that if the sum of $(4q+2)_s$ and $(4r)_a$ components is even then the reaction is photochemically allowed. If the number is odd and the process is observed under photochemical conditions, the reaction is not expected to be concerted. The generalised Woodward–Hoffmann rule has application to many of the reaction types discussed in this text and it will be used in the relevant discussions.

For any electrocyclic reaction there are two conrotatory and two disrotatory modes of ring cleavage and ring closure. The two conrotatory modes for the ring opening of *cis*-3,4-dimethylcyclobutene (9) lead to the

Fig. 2.6. The buta-1,3-diene-cyclobutene interconversion (a) with conrotation and (b) with disrotation.

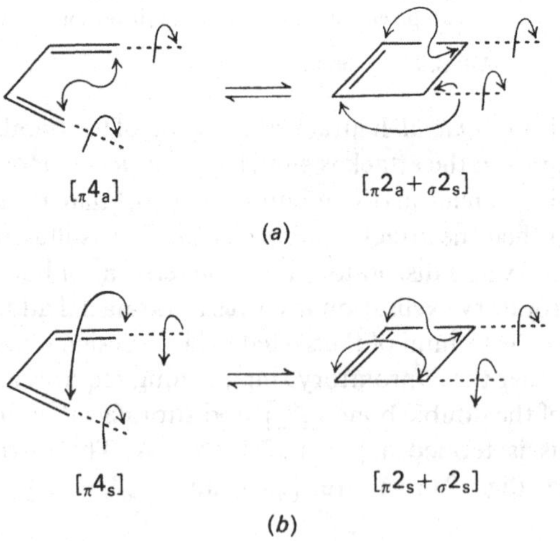

$[_\pi 4_a]$ $[_\pi 2_a + _\sigma 2_s]$

(a)

$[_\pi 4_s]$ $[_\pi 2_s + _\sigma 2_s]$

(b)

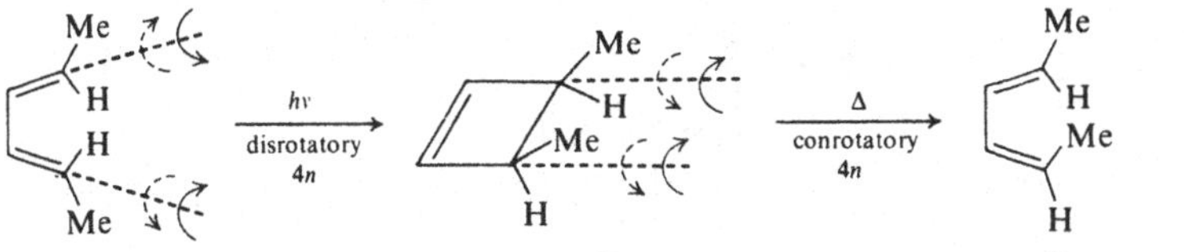

cis-(9)   (10)

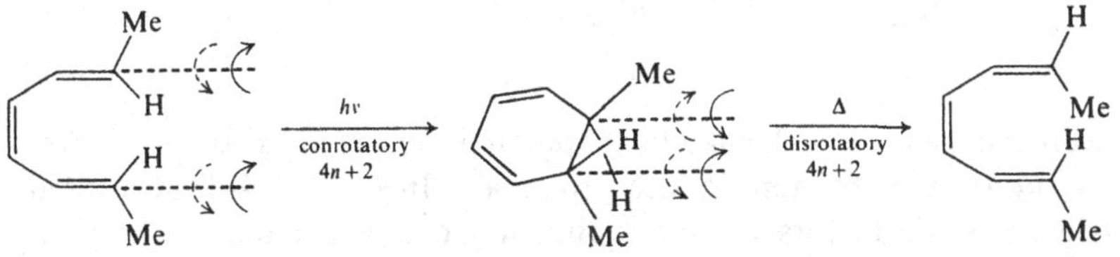

trans-(11)

same product (10); the same is true for the two disrotatory modes for cleavage of *trans*-5,6-dimethylcyclohexa-1,3-diene (11). On the other hand disrotatory photochemical opening of *cis*-3,4-dimethylcyclobutene can lead to two products, *cis,cis*- and *trans,trans*-hexa-2,4-diene (12). In like

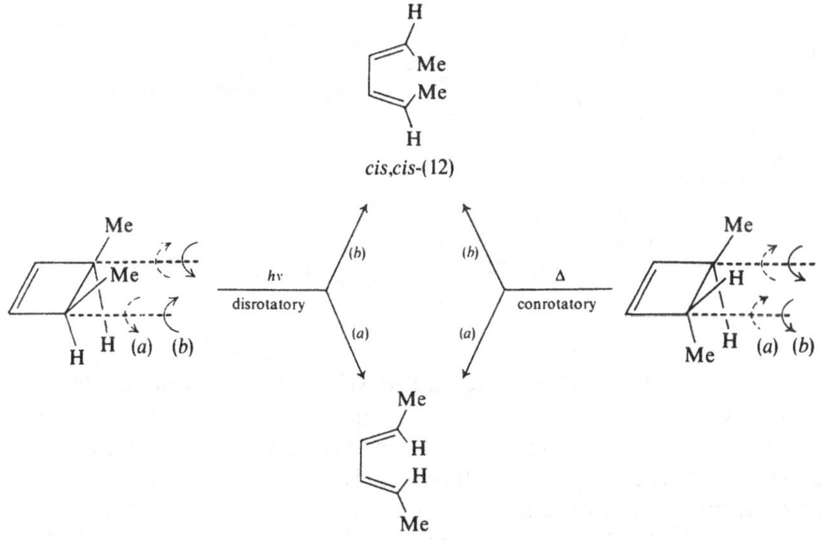

cis,cis-(12)

trans,trans-(12)

manner thermal conrotatory opening of the *trans*-dimethylcyclobutene can lead to two products (12). In the thermal reaction only the *trans,trans*-diene (route (*a*)) is obtained. The exclusive formation of *trans, trans*-hexa-2,4-diene results from steric interference to the two methyl groups in the transition structure leading to the *cis,cis*-diene isomer (route (*b*)).

For the electrocyclic closure of a diene to proceed it is essential that the ss-*trans*   s-*cis*

s-*trans*   s-*cis*

cis conformation is adequately populated since it is only in this conformation that the termini are proximate. In *cis,cis*-1,4-disubstituted derivatives steric factors disfavour the s-*cis* conformer and as a result electrocyclic closure of such systems is rarely observed. The isomeric *cis, trans* derivatives still contain significant congestion in the s-*cis* form and only when terminal substituents are small is electrocyclic closure observed. For example, *cis*-penta-1,3-diene cyclises to 3-methylcyclobutene (13) but the quantum yield is ten times smaller than for the *trans*-isomer. On the other hand *trans,trans*-1,4-disubstituted buta-1,3-dienes have minimal

trans-   (13)   *cis*-

interference in the s-*cis* form and electrocyclic closure is facilitated; *trans, trans*-hexa-2,4-diene is the only hexa-2,4-diene isomer to undergo closure.

For the electrocyclic reaction of a triene to occur the central double bond must have a *cis*-configuration otherwise the termini of the triene cannot come within bonding distance. The influence of ground state conformation, particularly in the closure of acyclic trienes, is more subtle.[15] Substitution at the 2- and 5-positions of the triene favours the s-*cis*,s-*cis* conformation and irradiation of these substrates gives the cyclohexadiene as a major product of reaction. However, the alternative electrocyclic reaction, namely cyclobutene formation from the s-*cis*,s-*trans* form, is also observed (scheme 2.4). In many polyenes (or polyfunctional molecules) more than one electrocyclic (or pericyclic) reaction path is possible. While the Woodward–Hoffmann approach does not in itself distinguish which of the allowed

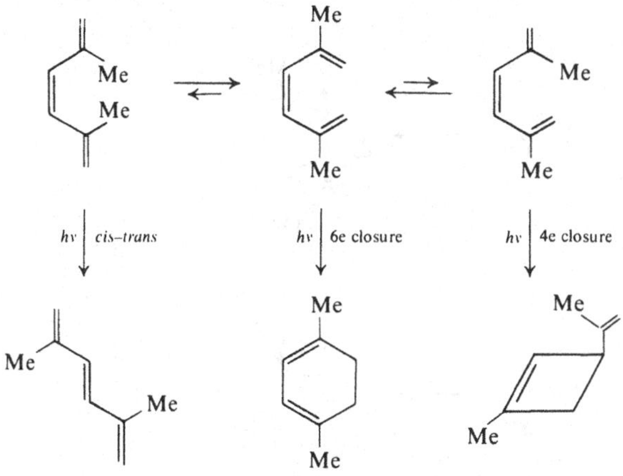

Scheme 2.4

electrocyclic (or pericyclic) reactions will occur, extensions of the frontier orbital approach using *perturbation molecular orbital* (PMO) theory often allow comment to be made on *periselectivity – the selection of one orbital symmetry allowed pathway over another* – as well as *regioselectivity* (see section 2.3).[9-14]

Because of conformational constraints it is not surprising that many of the most quoted examples of photochemical electrocyclic reactions involve cyclic systems where the conjugated alkene is held with its termini within bonding distance. The examples shown on the following pages, selected from carbocyclic and heterocyclic chemistry, illustrate the principles of the Woodward–Hoffmann symmetry rules for electrocyclic reactions and suggest something of the scope these reactions have in synthesis.

Electrocyclic reactions of trienes contained in seven-membered rings are well known. Cycloheptatriene itself affords bicyclo[3,2,0]hepta-2,6-diene (14) by disrotatory closure of a 4π-system. The alternative allowed electrocyclic closure across the termini of the 6π-system would be expected

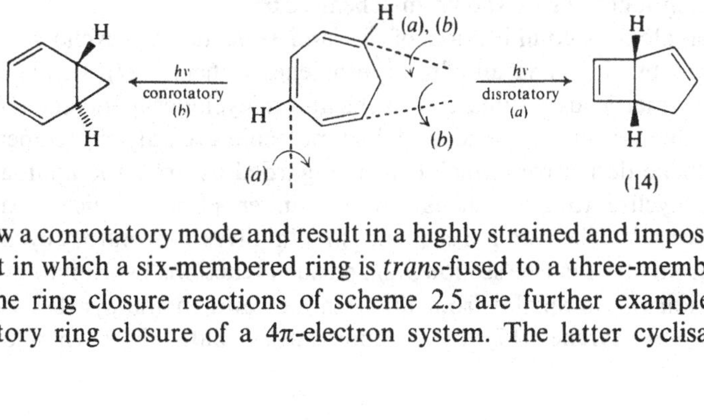

to follow a conrotatory mode and result in a highly strained and impossible product in which a six-membered ring is *trans*-fused to a three-membered one. The ring closure reactions of scheme 2.5 are further examples of disrotatory ring closure of a 4π-electron system. The latter cyclisation

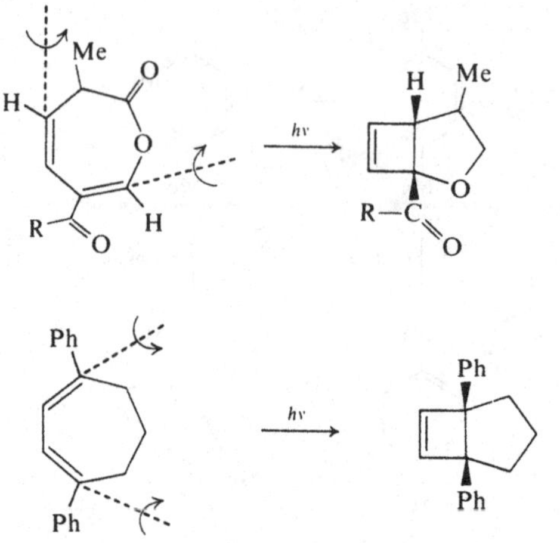

Scheme 2.5

proceeds smoothly despite the steric interference of the phenyl substituents in the product.

The *cis,cis*-bicyclic diene (15) undergoes symmetry-allowed photochemical conrotatory ring opening to give *cis,cis,trans*-cyclonona-1,3,5-triene, a

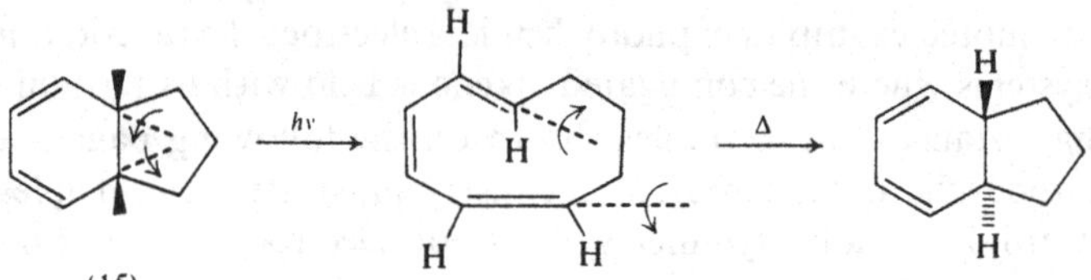

(15)

molecule which at room temperature undergoes disrotatory closure to give the *trans*-fused isomer of the starting material. The *cis,cis*- and *cis,trans*-cyclodeca-1,3-dienes undergo thermal conrotatory and photochemical disrotatory ring closure to yield the bicyclo[6,2,0]decenes with *cis*- and *trans*-ring junctions as shown in scheme 2.6.

An analogous equilibration of *cis*- and *trans*-fused bicyclic alkenes is provided by the 4a,8a-dihydronaphthalenes (scheme 2.7). Irradiation of either isomer leads (ultimately) to the all-*cis*, cyclodecapentaene (16) as a thermally and photochemically labile molecule even at low temperature. *cis*-Stilbene derivatives which can be regarded as aromatic analogues of trienes cyclise to give phenanthrenes under photochemical oxidative conditions.[16] The reaction sequence first involves conrotatory cyclisation and then oxidation (see section 5.2) and such reactions have analogy in the formation of carbazoles from diarylamines and in the synthesis of the substituted pyrene (17) by a conrotatory path. The thermal and

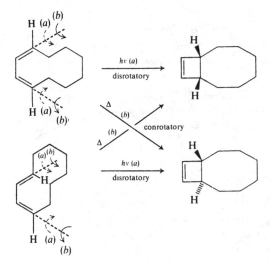

Scheme 2.6

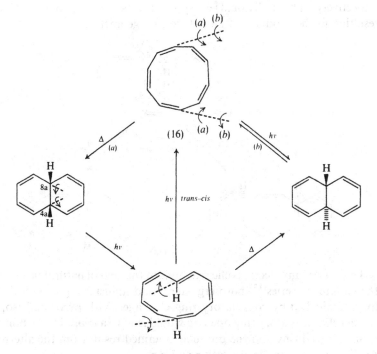

Scheme 2.7

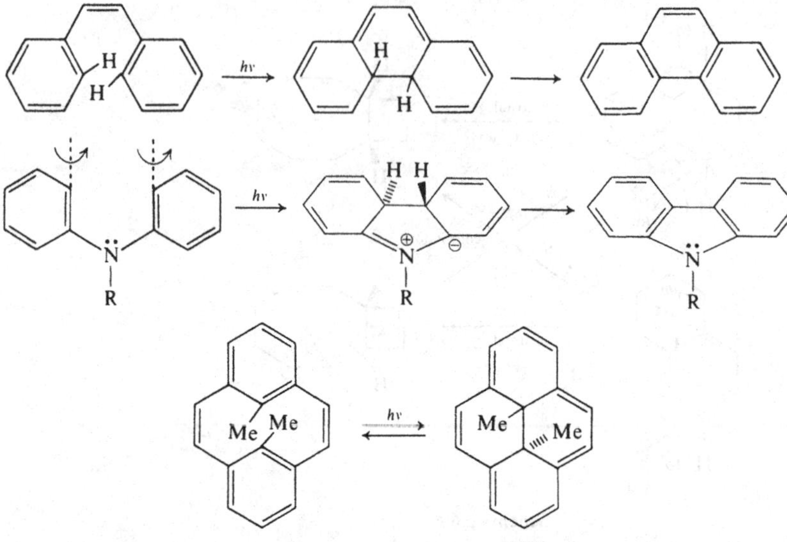

(17)

photochemical isomerisation of [16]annulene lead to the cyclooctadiene derivatives (18) by double disrotatory and conrotatory processes respectively. The individual 6π-systems close in the opposite sense, resulting in the products of most favoured geometry.

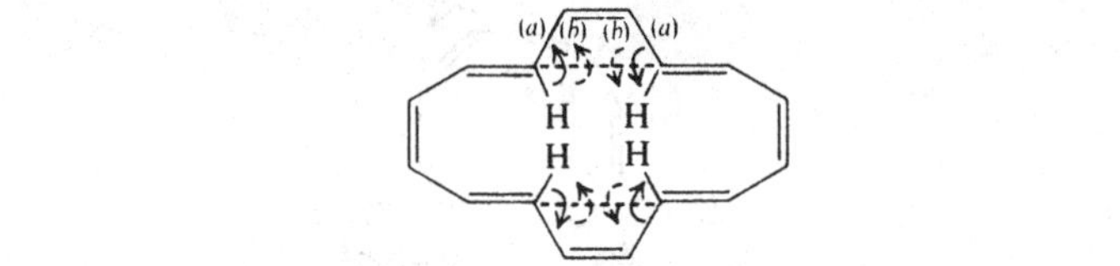

(18)

There are many electrocyclic reactions in the field of natural products. In the vitamin D series[17] both ergosterol and lumisterol give precalciferol (previtamin D_2) by conrotatory ring cleavage. With pyro- and isopyro-calciferol, conrotatory ring opening would lead to a *trans* double bond in a six-membered ring and the products obtained result from the alternative 4π-disrotatory ring closure (scheme 2.8).

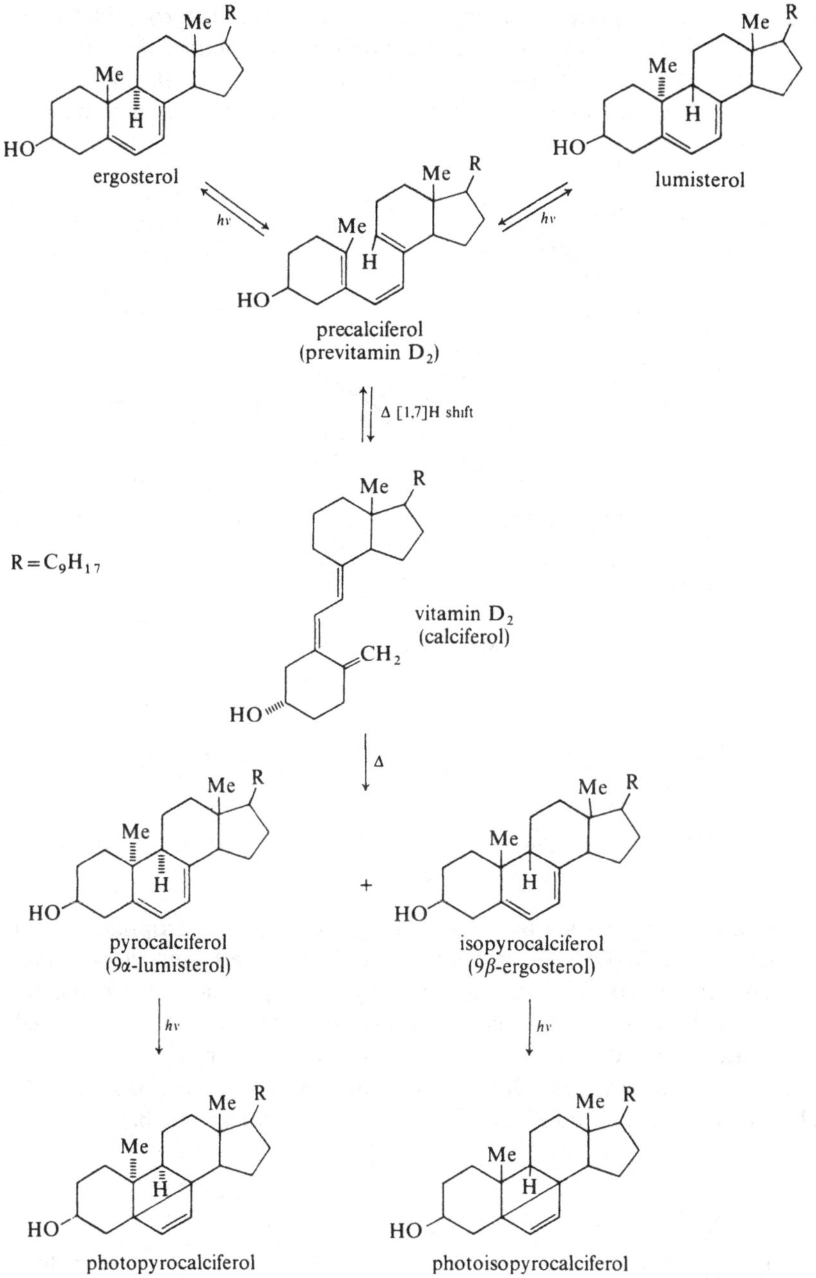

R = C$_9$H$_{17}$

Scheme 2.8

Because of the special strained nature of σ-bonds in heterocyclic three-membered ring derivatives and the presence of a non-bonding electron pair, these compounds undergo four electron disrotatory photochemical cleavage. Thus the isomeric aziridines (scheme 2.9) undergo electrocyclic

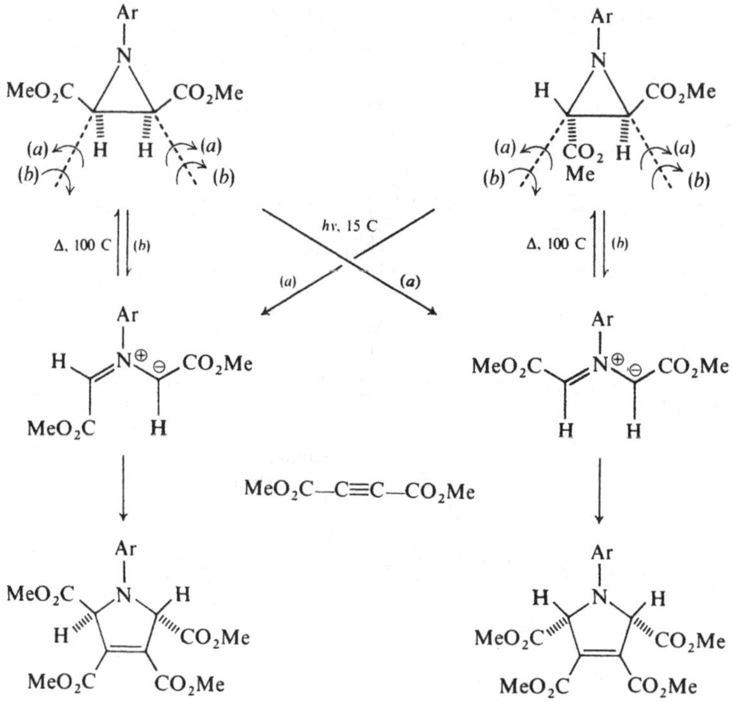

Scheme 2.9

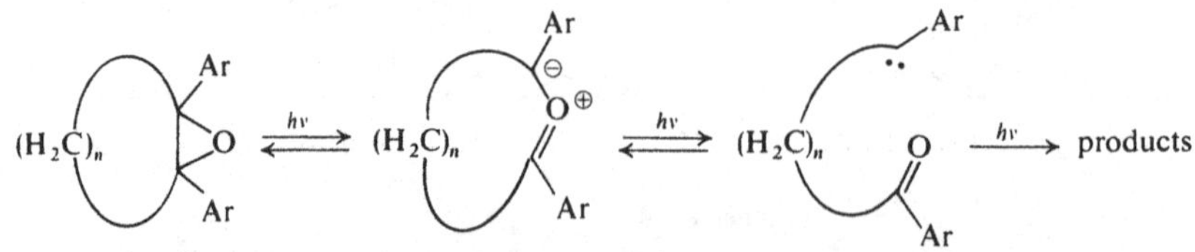

reaction to produce the open-chain 4π-electron 1,3-dipolarophiles which can be trapped by Diels–Alder addition to dimethyl ethynedicarboxylate. The differing stereochemistries of the Δ^3-pyrroline products obtained from the thermal and photochemical reactions are consistent only with initial conrotatory and disrotatory aziridine ring cleavage respectively. With oxiranes, carbonyl ylides (19) are formed on photolysis, and while the ylide (19; $n=2$) is stable – and is responsible for the photochromic behaviour of cyclobutene oxide – those from cyclopentene- and cyclohexene-oxides are converted into carbenes. The photolysis of thirians results in sulphur extrusion and an electrocyclic pathway is unlikely.

Several highly strained molecules have been synthesised by routes involving electrocyclic ring closures. For example, bicyclo[2,1,0]pent-2-ene

(20) $Y = CH_2$, NR,
O, S

results from photolysis of cyclopentadiene (20; $Y = CH_2$). In fact, the reaction is general for the six-electron aromatic heterocycles (20) and proceeds from the S_1 excited state with disrotation. However, the bicyclic product is usually thermally or photochemically labile and has only been isolated for the parent system when $Y = CH_2$ and from thiophene derivatives (20; $Y = S$). Cyclobutadiene iron tricarbonyl has been formed by photochemical decarboxylation of the cyclisation product of the unsaturated lactone shown, and the application of photochemical methods

to such organometallic synthesis is becoming more important. A similar reaction sequence has facilitated the synthesis of bicyclo[2,2,0]hexadiene or 'Dewar benzene'. This historically interesting molecule is highly strained but, even so, it has a half-life of about two days in pyridine solution at room temperature. The relative stability of Dewar benzene has an interesting origin; the thermally-allowed conrotatory ring opening is unfavourable because of the highly strained nature of the resultant *cis,cis,trans*-triene. Ring opening to benzene does occur, however, probably by cleavage to a diradical.

Dewar benzene can also be obtained from the solution-phase photolysis of benzene at short wavelength.[18] Irradiation of benzene at 204 nm causes S_0–S_2 excitation and calculations indicate that bonding between the *para*-positions is a favourable process in this state; Dewar benzene is formed by the equivalent of a 4π-electron disrotatory process. In the S_1 excited state (irradiation at 254 nm) it is the *meta*-positions that are incipiently bonding

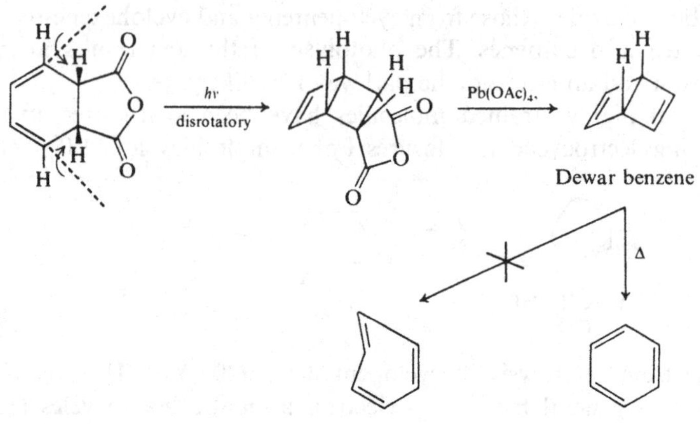

Dewar benzene

and benzvalene (21) and fulvene (22) are produced (see section 4.5). Benzvalene is also produced from the S_2 state, but Dewar benzene is *not* produced from the S_1 excited state. Furthermore, Dewar benzene has the

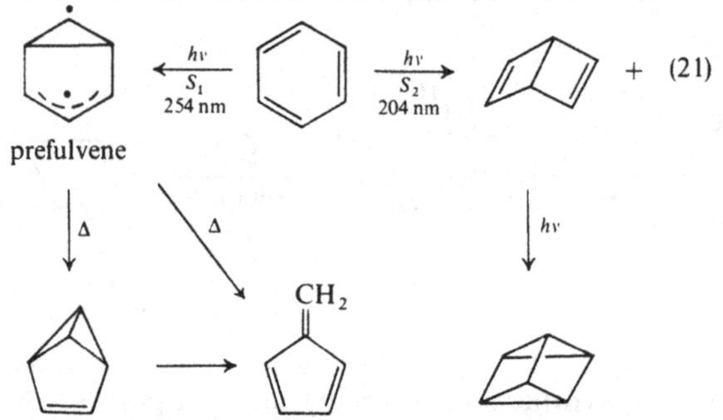

prefulvene

benzvalene   fulvene   prismane
(21)   (22)   (23)

capacity to undergo a photochemically-allowed intramolecular $[_\pi 2 + _\pi 2]$ cycloaddition to deliver the valence bond isomer prismane. The photolysis of alkyl substituted benzenes takes advantage of the various pathways to provide one or more of the valence bond isomers or a positional isomer of the starting material as illustrated for 1,2,4-tri-*t*-butylbenzene in scheme 2.10. The formation of the 1,3,5-isomer results from cycloreversion of the prismane depicted or arises via the benzvalenes to yield the least sterically congested product. By an analogous sequence *ortho*-xylene can give the *meta*- and *para*-isomers. 'Dewar pyridine' has been isolated from irradiation of pyridine in the liquid phase and the molecule has a half-life of 36 minutes at 0°C.

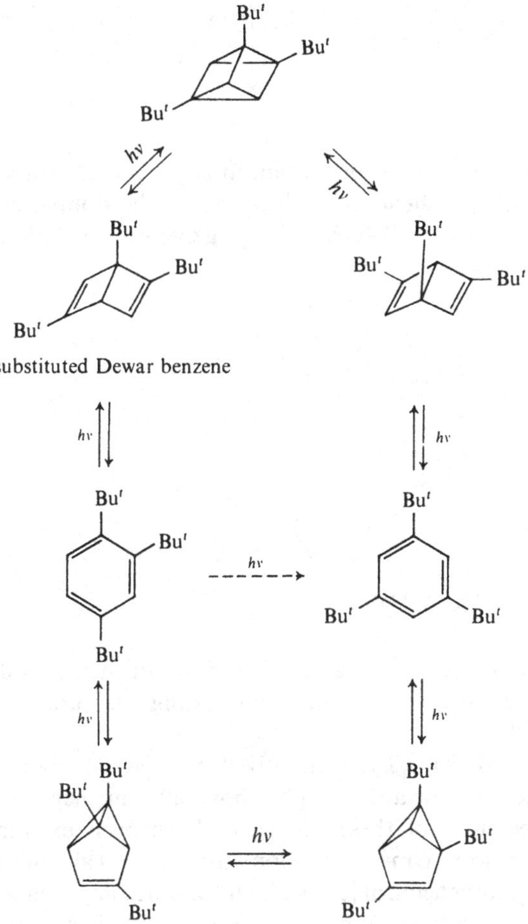

substituted Dewar benzene

Scheme 2.10

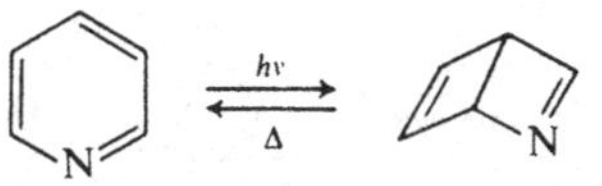

Dewar pyridine

2.3 Rearrangements – 1,4-dienes and the di-π-methane or Zimmerman rearrangement[19,20]

The commonest photochemical reaction of 1,4-dienes is the *di-π-methane or Zimmerman rearrangement*, in which the 1,4-diene is converted into a vinylcyclopropane (24). A 'di-π-methane' unit involves a saturated carbon atom directly attached to two sites of unsaturation, as exemplified

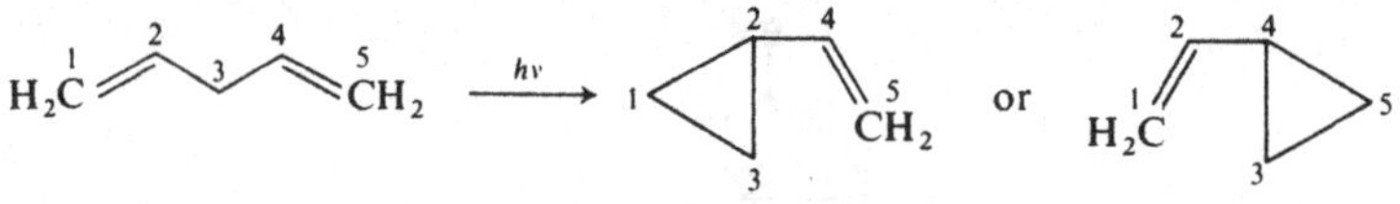

(24)

by penta-1,4-diene above. The transformation is shown schematically in scheme 2.11 for penta-1,4-diene and when one of the double bonds is replaced by an aromatic ring. The reaction is general for solution-phase

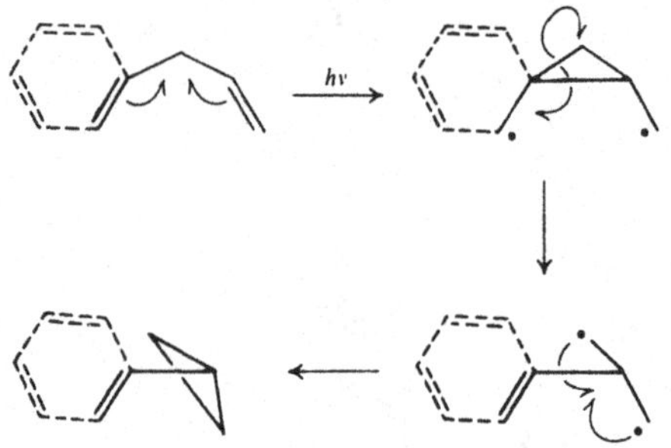

Scheme 2.11

photolysis of acyclic-, cyclic-, bi- and tri-cyclic-1,4-dienes and detailed investigations have revealed many interesting features of the rearrangement.[19,20]

Penta-1,4-diene and its alkyl derivatives possess isolated alkenic chromophores which do not absorb light above 200 nm; hence only the triplet-sensitised reactions of these molecules have been examined. In addition to vinylcyclopropane formation the diene can undergo a symmetry-allowed photochemical $[_\pi 2_s + _\pi 2_s]$ intramolecular cycloaddition to give a bicyclo[2,1,0]pentane. Direct photolysis of a 1,4-diene can only be

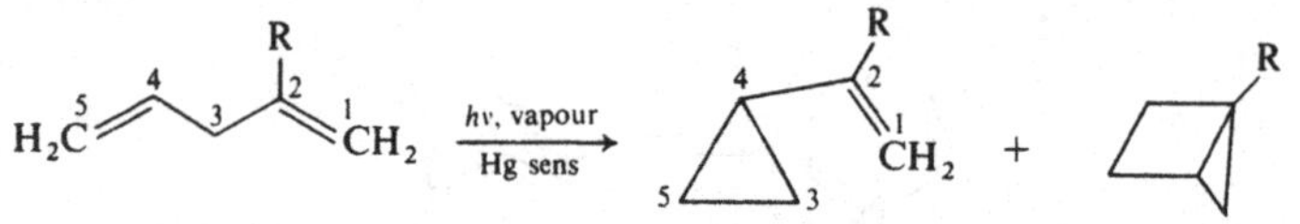

examined if one of the double bonds has extended conjugation since this shifts absorption into the accessible region of the ultraviolet spectrum. Direct photolysis of 3,3-dimethyl-1,1,5,5-tetraphenylpenta-1,4-diene (25) has been found to give the vinylcyclopropane (26) as the sole primary photoproduct. With different substituents at C_1 and C_5, as shown in scheme 2.12, two reaction routes (path (*a*) and path (*b*)) become possible. However for this substrate the rearrangement occurs exclusively by path (*a*) to give 1,1-dimethyl-2,2-diphenyl-3(2,2-dimethyl)vinylcyclopropane (28).

Me 3 Me
2 4
Ph 1 5 Ph
Ph Ph
(25)

$\xrightarrow{h\nu}_{\pi \to \pi^*}$

Me 3 Me
1 2 4
Ph 1 5 Ph
Ph Ph

Me 3 Me
2 4
Ph 1 5 Ph
Ph Ph

←

Me 3 Me
2 4
Ph 1 5 Ph
Ph Ph

Me 3 Me
2 4
Ph 1
Ph Ph 5 Ph
(26)

This *regiospecific*† process results from cleavage of the three-membered ring of diradical (27) towards the isopropyl radical (path (a)) rather than towards the more stable diphenylmethyl radical (path (b)). Regioselectivity is observed for the diaryldiphenyldienes (29a,b). For each substrate, cleavage of the cyclopropane diradical occurs towards what is considered to be the more electron-rich centre. That is for diene (29a; Ar = p-MeO–C$_6$H$_4$) path (a) involving rupture towards the substituted diarylmethyl centre dominates while for diene (29b; Ar = p-NC–C$_6$H$_4$) cleavage towards the diphenymethyl centre (path (b)) occurs. These observations are supported by molecular orbital calculations which indicate that C$_1$ and C$_5$ are electron rich in the singlet excited state and that opening of the three-membered ring towards the most electron-rich centre is favoured. Nonetheless, kinetic studies have shown that (29a), with p-methoxy substitution undergoes singlet state rearrangement at a rate ($k = 3.4 \times 10^9$ s^{-1}) slower than the unsubstituted compound (25) ($k = 1.4 \times 10^{11}$ s^{-1}). With *para*-cyano substituents a slight rate enhancement ($k = 2.2 \times 10^{11}$ s^{-1}) is recorded. The calculations suggest that the electron-donor substituents

† *Regio specificity* may be defined as exclusive reaction at one region of the molecule as opposed to an alternative region.

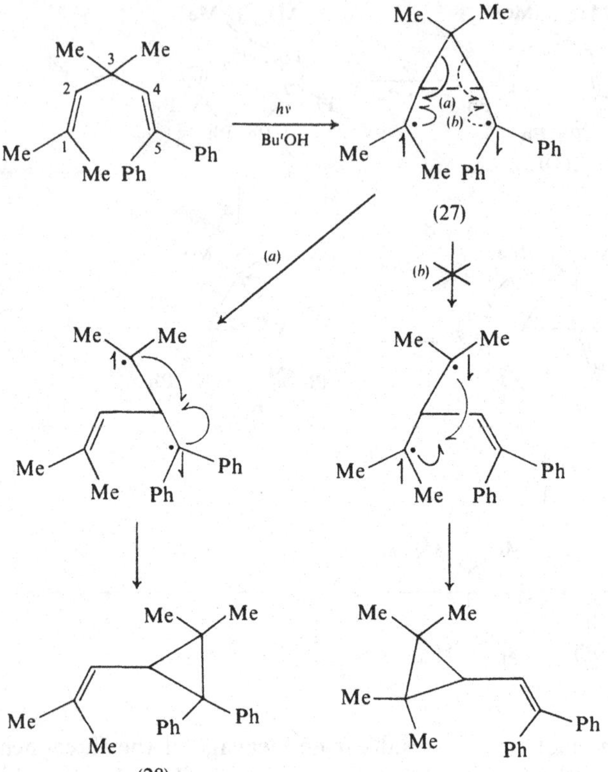

Scheme 2.12

decrease the rate of reaction by stabilising the initially produced excited state in its high vibrational level (vertical excitation), a stabilisation which exceeds that available for the diradical formed in the rate limiting step.

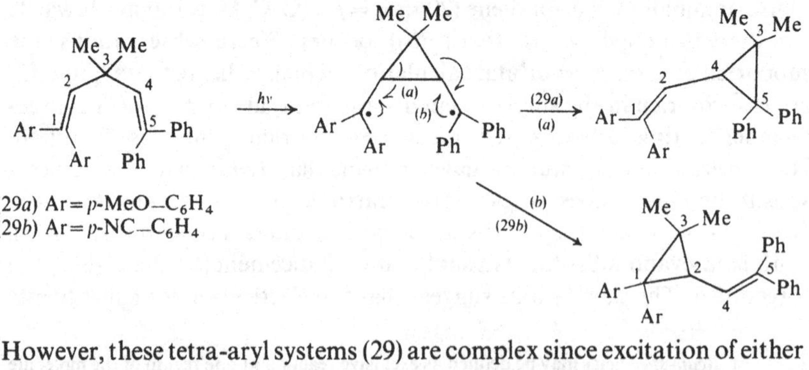

(29a) Ar = *p*-MeO—C_6H_4
(29b) Ar = *p*-NC—C_6H_4

However, these tetra-aryl systems (29) are complex since excitation of either conjugated chromophore can occur and a substituent will influence not only the energy required for absorption, but also the extent to which the excited chromophore interacts with its proximate ground state counterpart. Furthermore, a substituent will have an effect on the reaction following cyclopropyl bond formation.

The aryl substituted dienes thus far discussed undergo Zimmerman rearrangement in the singlet excited state but are recovered unchanged when subjected to sensitised photolysis. However, when one of the alkenic groups carries dissimilar terminal substituents, triplet sensitisation results in observable geometrical isomerisation (scheme 2.13); rearrangement

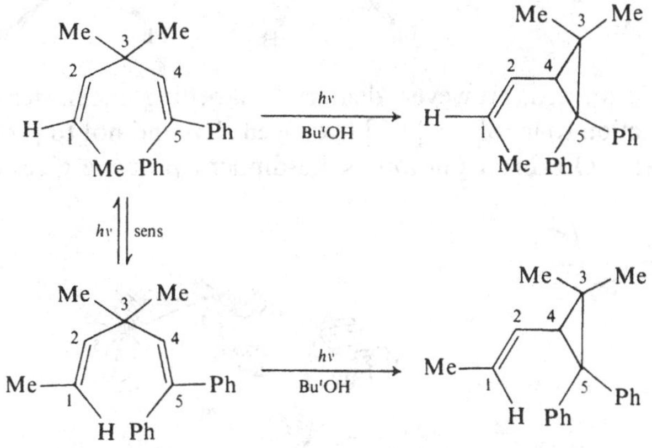

Scheme 2.13

occurs on direct irradiation, but without alkene isomerisation. Thus reaction from the S_1 state is both regiospecific and stereospecific. These latter observations place a limit on the lifetime of the cyclopropyldicarbinyl diradical since rotation about the C_1–C_2 bond does not compete with vinylcyclopropane formation.

The photolysis of the optically active substrates with chirality at C_3 results in the formation of geometrical isomers of the cyclopropane each of which has been shown by careful correlation studies, to have undergone inversion of configuration at C_3. These reactions define the mode of attack at each of C_1, C_3, and C_5 and it is worthy of note that the reaction can be considered as a photochemically-allowed cycloaddition with the notation $[_\pi2_s + _\sigma2_a + _\pi2_s]$. The reaction in general involves both double bonds and cannot be regarded as a $[_\pi2_a + _\sigma2_a]$ reaction since the corresponding pentene does not undergo the di-π-methane rearrangement. Zimmerman chooses to describe the di-π-methane rearrangement as a Möbius array which corresponds to a delocalised transition structure for an allowed photochemical, but not ground state reaction.

In the absence of the *gem*-dimethyl substitution at C_3, a di-π-methane

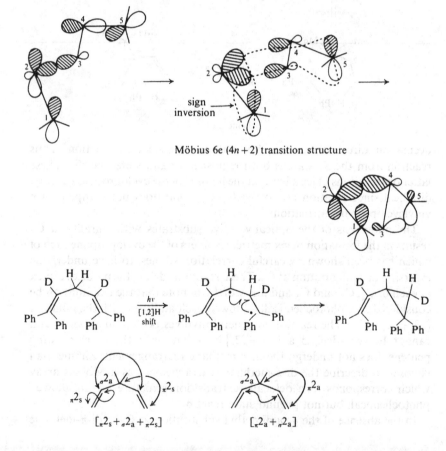

product is still formed. However, deuterium labelling experiments have shown the reaction to involve a [1,2] hydrogen shift and not to proceed in the normal way. On direct photolysis 1,3-diphenylpropene gives *cis*-1,2-

Möbius 6e $(4n+2)$ transition structure

diphenylcyclopropane. The absence of deuterium migration in labelled substrates establishes the route to be by the normal di-π-methane pathway.

A large number of rearrangements can be classified as of di-π-methane type even though the molecules do not formally contain a 1,4-diene unit. For example, as we have already indicated, an aryl ring may replace one of the alkenic bonds with subsequent benzo-vinyl bonding and migration of this aryl group. The *gem*-dimethyl functionality is not essential for the aryl–vinyl rearrangement to occur. As might be expected the presence of substituents capable of stabilising the radical facilitate reaction. Thus for (30*b*; X = OMe) the *para*-methoxy group increases the rate of reaction by a

(30*a*) X = H
(30*b*) X = OMe
(30*c*) X = CN

factor of 5.2, whereas for (30*c*; X = CN) the enhancement is only 1.2. On replacing the diphenylvinyl group (Ph$_2$C=C) of (30) by a styryl moiety (PhCH=C) both the singlet state di-π-methane reaction and IC into the ground state are markedly slowed. This contrasts with the tetra-arylpenta-1,4-dienes (29) where *para*-methoxy substitution slows the rate and substitution on the methane carbon is necessary for the di-π-methane rearrangement to occur.

For (*E*)-1-aryl-3-methyl-3-phenylbut-1-enes (31) (scheme 2.14) the migrating group in each case is phenyl and therefore the substituent effects might be expected to be easier to unravel. However, these compounds undergo (*E*)–(*Z*) isomerisation as well as the di-π-methane rearrangement. The relative amount of (*E*)–(*Z*) isomerisation is increased by *para*-methoxy, *meta*-chloro, 3,4-dichloro and *para*-methyl substituents and this could reflect the ability of these groups to enhance ISC to the T_1. The presence of *para*-methoxy and 3,4-dichloro groups also inhibits the formation of the cyclopropane. The arylphenylcyclopropane products of rearrangement are themselves photolabile and on extended photolysis undergo a cyclopropyl–allyl rearrangement (scheme 2.14). Cleavage of the ring between the dimethyl substituted carbon atom C_1 and the benzylic

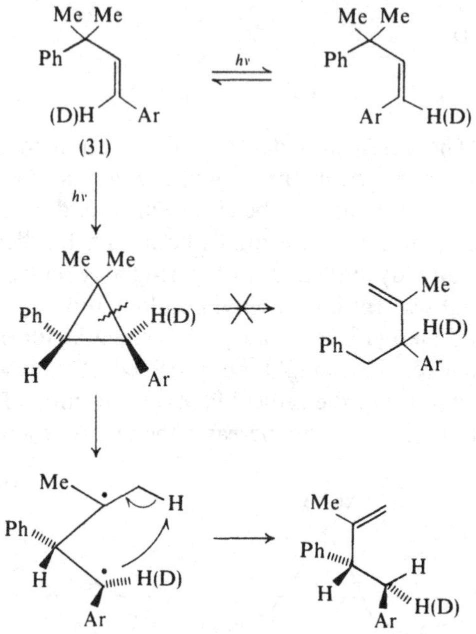

Scheme 2.14

carbon adjacent to the aryl group (*para*-Me, *para*- and *meta*-Cl) occurs as
shown in scheme 2.14.

The di-π-methane rearrangement proceeds by a singlet mechanism for
acyclic and monocyclic systems whilst the triplet sensitised reactions of the

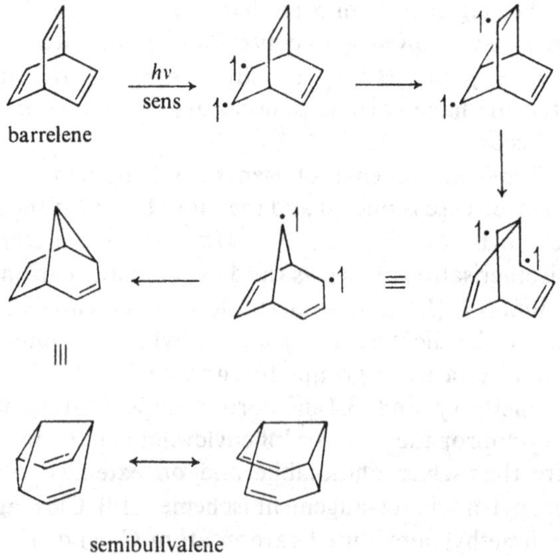

dienes usually result in geometrical isomerisation. However, the di-π-methane rearrangement is not confined to acyclic and monocyclic systems. Bicyclic-1,4-dienes also rearrange but by way of a triplet and not a singlet process; in all probability, the in-built geometric constraint of a bicyclic molecule does not allow for distortion of the triplet excited state. For example, barralene (bicyclo[2,2,2]octatriene) rearranges to semibullvalene, a molecule which undergoes rapid valence bond tautomerism. The sensitised reaction is thought to proceed in a stepwise fashion unlike the acyclic singlet case which can be a concerted symmetry-allowed reaction. Benzbarrelene undergoes the di-π-methane rearrangement on sensitised photolysis but gives benzcycloocta tetraene on direct irradiation. Deuteration studies have shown (scheme 2.15) that the benzcyclooctatetraene is formed to an extent of 94% by route (b) and the

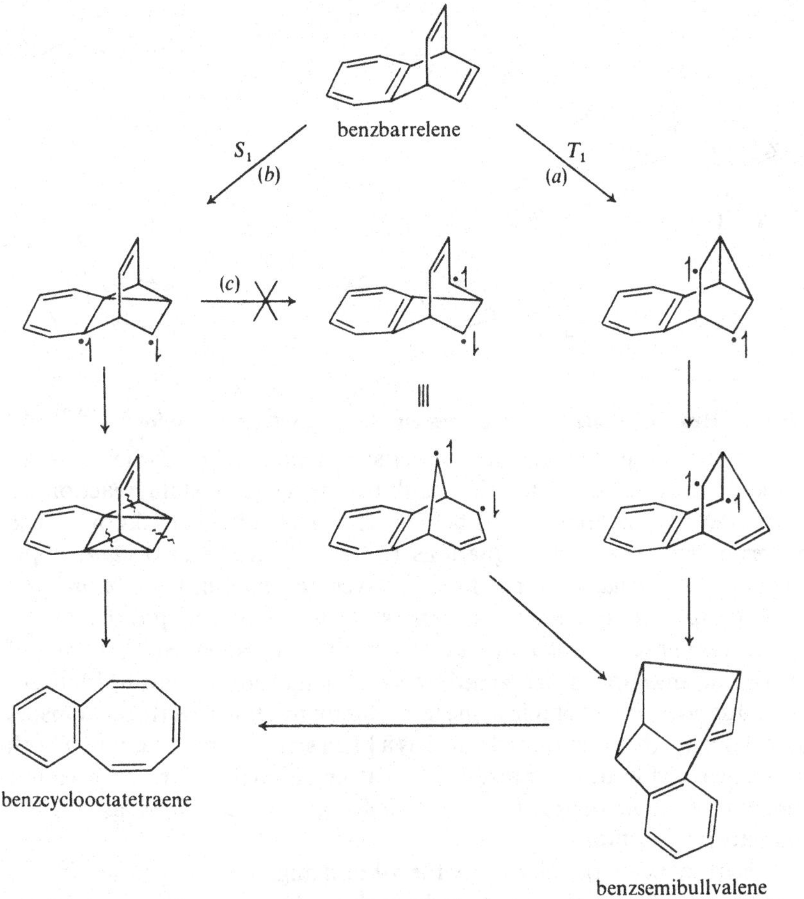

Scheme 2.15

benzsemibullvalene entirely by route (*a*). The remainder of the benzcyclooctatetraene (6%) comes via the benzsemibullvalene. Direct irradiation of benzbarrelene therefore proceeds almost entirely by benzo–vinyl overlap and the cycloaddition is probably an allowed $[_\pi 2 +_\pi 2]$ process. On the other hand, the sensitised reaction proceeds by route (*a*) which involves vinyl–vinyl overlap with no benzsemibullvalene being formed via route (*c*) involving benzo–vinyl overlap. Where there is a choice, vinyl–vinyl bonding is preferred since there is no disruption to aromaticity.

Benzbicyclo[2,2,1]heptenes, e.g. (32), also yield di-π-methane products on sensitised irradiation. With a vinylcyanide the cyano group dictates the triplet pathway so that bridging yields the cyano stabilised radical.

(32)

X = H or OMe

2.4   Rearrangements – 1,5-dienes and the sigmatropic reaction[8−14,21−23]

The excited singlet and triplet state reactions of 1,5-dienes are in marked contrast to each other and to the ground state reaction as illustrated by reactions (i)–(iii) below each of which is well known. The processes represented by equations (i) and (ii) are termed *sigmatropic* reactions. A *sigmatropic* reaction involves *the migration of a σ-bond, adjacent to one or more π-electron systems, to a new position in an uncatalysed intramolecular reaction.*[8−14,21] The so-called reaction order of such a rearrangement is determined by counting the atoms across which the σ-bond migrates. For example if one terminus migrates across three atoms, as in reaction (i), the reaction is defined as a [1,3] sigmatropic reaction. If both termini have moved across three carbon atoms, as in (ii), the reaction can be thought of as involving two [1,3] shifts and is described as a [3,3] sigmatropic reaction.

The group being transferred by the σ-bond migration may be associated with the same face of the π-system throughout. In this case the migration is

(i) [1,3] sigmatropic reaction

(ii) Cope rearrangement: a [3,3] sigmatropic reaction

(iii) 1,5:2,6 cross-cycloaddition

referred to as a *suprafacial* sigmatropic shift. This is illustrated by the photochemically induced degenerate [1,3] suprafacial shift of H_A in propene. If the group being transferred passes from one face of the π-system to the opposite face, the process is referred to as an *antarafacial* sigmatropic shift. This is illustrated by the thermal [1,7] shift of H_A across hepta-1,3,5-

[1,3] suprafacial sigmatropic shift

[1,7] antarafacial sigmatropic shift

triene. With a longer conjugated system the molecule can twist to form a spiral thereby allowing the transfer of H_A from one face to the other in a concerted process. The spiral conformation allows for overlap of the molecular orbitals involved at all stages of the reaction. A similar process occurs for vitamin D_2 (see section 2.2).

Sigmatropic reactions are found to occur under both thermal and photochemical conditions and the application of orbital symmetry arguments allows the various shifts to be defined as *allowed* or *not allowed* and this has become synonymous with 'expected' or 'not expected'.

A [1,3] hydrogen shift can be formally treated as the migration of a hydrogen atom across an allyl radical or a proton across an allylic anion. The molecular orbitals of the allylic system are shown in fig. 2.7 as overlapping atomic p-orbitals. For conjugated systems with an even number of carbon atoms, such as butadiene or hexatriene, the nodes occur between the carbon atoms. However, with an odd number of carbon atoms the nodes can coincide with one or more of the carbon atoms and this occurs for the allylic system.

In the frontier orbital approach to a [1,3] shift the bond joining the migrating group and the polyene is considered (hypothetically) to break heterolytically. The LUMO of the migrating group, considered for this purpose as a cation, and the HOMO of the complementary allylic anion must be capable of overlapping so as to allow continuity in bond cleavage and bond formation. This is shown for the thermal reaction in fig. 2.8(*a*) where the HOMO of the anion is the ψ_2-orbital and the LUMO of the migrating fragment the hydrogen 1s-orbital. Suprafacial migration is not allowed, and even though antarafacial migration is allowed the steric constraint of such a process would not allow for bond breaking and bond formation to be synchronous. For the excited state reaction (fig. 2.8(*b*)) the HOMO of the hypothetical allylic fragment, the ψ_3-orbital, and the LUMO of hydrogen, the 1s-orbital, are such that suprafacial migration is allowed. Antarafacial migration is not allowed nor is it sterically feasible.

Fig. 2.7. Molecular orbitals of the allyl system.

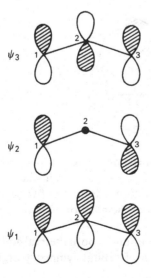

The transition state approach to a [1,3] sigmatropic shift of hydrogen employs atomic orbitals and the transition state structure is drawn as a series of overlapping s- and p-orbitals with phases inserted so as to minimise the number of phase changes in the participating atomic orbitals as depicted by (33). There are no phase inversions and so the array is

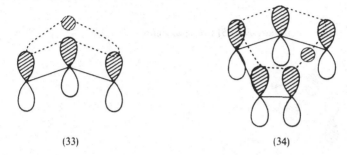

(33)   (34)

classified as Hückel. Since four electrons are involved in the transition, the reaction is antiaromatic and forbidden from the ground state but allowed from the first excited state (see table 2.3).

For a [1,5] hydrogen shift, the thermal suprafacial process is controlled by the HOMO ψ_3 of the hypothetical pentadienylic anion and the LUMO 1s-orbital of the proton (fig. 2.9). The migration is allowed. The photochemical shift, controlled by the HOMO, ψ_4, and the 1s-orbital is forbidden (fig. 2.9). For an antarafacial shift the thermal process is forbidden and the photochemical one is allowed. However, antarafacial migration would require the transition state structure to be so distorted that the reaction is not expected to be concerted. The suprafacial [1,5] sigmatropic shift of hydrogen possesses a Hückel atomic orbital transition state structure (34) since there are no phase inversions in the cyclic array. With the delocalisation of six electrons the reaction has an aromatic

Fig. 2.8. Frontier orbital approach to [1,3] hydrogen shifts (a) thermal (b) photochemical.

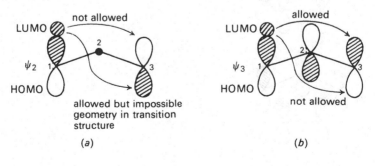

transition structure and is therefore a favoured process in the ground state (see table 2.3).

Unlike a hydrogen shift the sigmatropic migration of an alkyl group is complicated further since the group can migrate with either retention or inversion of configuration at the migrating centre. The reaction can also be analysed by the frontier orbital method with the overlapping 1s-orbital of

Fig. 2.9. Sigmatropic [1,5] hydrogen shifts.

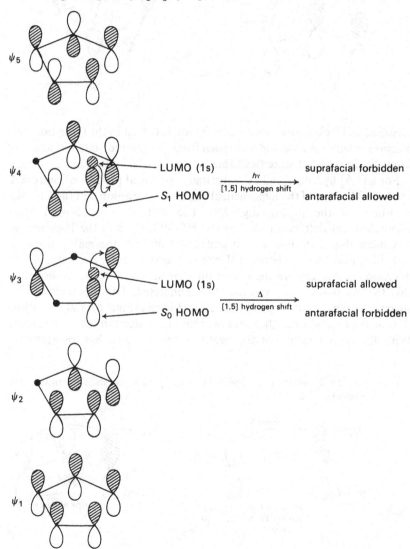

LUMO (1s)	*hv*	suprafacial forbidden
S_1 HOMO	[1,5] hydrogen shift	antarafacial allowed

LUMO (1s)	Δ	suprafacial allowed
S_0 HOMO	[1,5] hydrogen shift	antarafacial forbidden

hydrogen being replaced by the overlapping sp³-orbital of the migrating group. The photochemical [1,3] suprafacial shift of an alkyl group is allowed and proceeds with retention of configuration at the migrating carbon (fig. 2.10(*a*)) since the LUMO of the migrating fragment and the HOMO of the allylic anion are such as to allow continuity in bond cleavage and bond formation. A suprafacial [1,3] hydrogen shift is forbidden in the ground state, but a [1,3] carbon shift can occur. The reaction proceeds with inversion of configuration at the migrating carbon group and involves a 90° rotation to the transition structure and a further 90° rotation to give product as shown in fig. 2.10(*b*).

An alternative, but complementary, rationalisation of sigmatropic reactions can be achieved by considering the [1,3] carbon or hydrogen shift as an intramolecular cycloaddition between π- and σ-components. A favourable interaction between the HOMO–LUMO orbitals indicates a favourable energy barrier to reaction. The interactions between an occupied (HOMO) σ- or π-orbital and an unoccupied σ^*- or π^*-orbital respectively for the various stereochemical possibilities are shown in fig. 2.11 for the thermal and photochemical processes.

In the thermal reaction the HOMO–LUMO interaction is between either the σ- and π^*- or the σ^*- and π-molecular orbitals. The interactions of

Fig. 2.10. Suprafacial sigmatropic [1,3] allylic shift of carbon (*a*) photochemical (*b*) thermal.

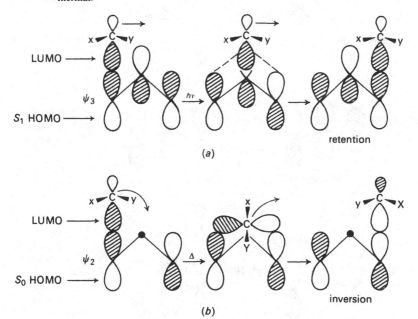

these orbitals are favourable when lobes of the same phase overlap in the development of the transition structure. The reaction diagrams show that this is the case for the $[_\pi 2_s + _\sigma 2_a]$ and $[_\pi 2_a + _\sigma 2_s]$ process (fig. 2.11(a)). Moreover, the processes can be shown from the Woodward–Hoffmann

Fig. 2.11. [1,3] Allylic shifts of carbon

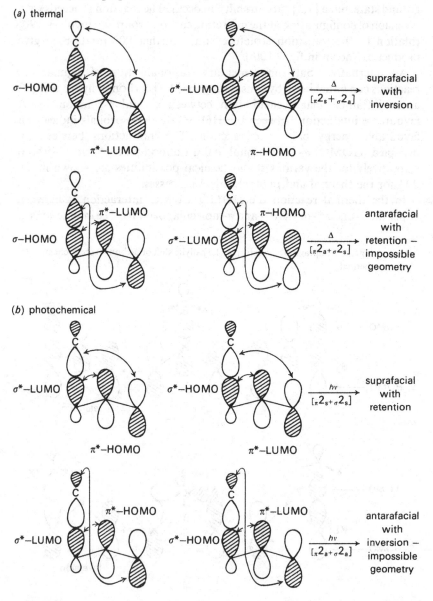

(a) thermal

σ–HOMO σ^*–LUMO $\xrightarrow[{[_\pi 2_s + _\sigma 2_a]}]{\Delta}$ suprafacial with inversion

π^*–LUMO π–HOMO

σ–HOMO π^*–LUMO σ^*–LUMO π–HOMO   antarafacial with retention – $\xrightarrow[{[_\pi 2_a + _\sigma 2_s]}]{\Delta}$ impossible geometry

(b) photochemical

σ^*–LUMO σ^*–HOMO $\xrightarrow[{[_\pi 2_s + _\sigma 2_s]}]{h\nu}$ suprafacial with retention

π^*–HOMO π^*–LUMO

σ^*–LUMO π^*–HOMO σ^*–HOMO π^*–LUMO   antarafacial with inversion – $\xrightarrow[{[_\pi 2_a + _\sigma 2_a]}]{h\nu}$ impossible geometry generalised rule to be allowed thermal reactions. For example, in the $[_\pi 2_s + _\sigma 2_a]$ reaction the $_\pi 2_s$ component is of the $(4q + 2)_s$ type $(q = 0)$ and there is no component of the $(4r)_a$ type. Thus the total number of $(4q + 2)_s$ and $(4r)_a$ components is *one* and the process is thermally allowed with inversion of configuration at the migrating centre. The alternative $[_\pi 2_a + _\sigma 2_s]$ mode of reaction involves antarafacial migration of the alkyl group with retention of configuration at the migrating centre. The reaction has one component of the type $(4q + 2)_s$ and $(4r)_a$, namely the $(_\sigma 2_s)$ component. The reaction is therefore orbital symmetry allowed. However, such a reaction would not be expected on steric grounds and, if observed at all, it could not be concerted since the transition structure would be too constrained for effective orbital overlap during the course of reaction; the reaction is said to be geometrically impossible at least as a concerted process.

In the photochemical reaction of this system an electron is promoted from the π- to the π^*-orbital and thus the HOMO–LUMO interaction will now be between π^* (HOMO) and σ^* (LUMO) (fig. 2.11(b)). The alternative combination involves the more energetic promotion of an electron from the σ- to the σ^*-molecular orbital and a σ^*(HOMO)–π^*(LUMO) interaction. However, the frontier orbital method shows that both orbital arrays are energetically favourable for a suprafacial [1,3] shift with retention of configuration. The reaction can be described as a $[_\pi 2_s + _\sigma 2_s]$ reaction. From the Woodward–Hoffmann generalised rule the reaction is allowed since the sum of the $(4q + 2)_s$ and $(4r)_a$ components is even; both the components of the reaction are of the $(4q + 2)_s$ type. The antarafacial migration of the alkyl group with inversion at the migrating centre is an allowed $[_\pi 2_a + _\sigma 2_a]$ process but the geometrical requirements do not allow for effective orbital overlap in the reaction.

The following two examples of thermal [1,3] suprafacial carbon shifts occur with inversion of configuration despite the presence of steric constraints against such inversion. The reactions demonstrate the importance of symmetry in dictating reaction course.

Table 2.4. *Orbital symmetry rules for* [1,j] *sigmatropic reactions*

Reaction	$1+j=4n$	$1+j=4n+2$
Δ	supra + inv.[1]	supra + ret.
	or antara + ret.	or antara + inv.[1]
hv	supra + ret.	supra + inv.[1]
	or antara + inv.[1]	or antara + ret.

[1] Inversion reactions cannot occur for sigmatropic hydrogen shifts.

Table 2.5. *Orbital symmetry rules for* [i,j] *sigmatropic reactions where i and j are greater than unity*[1]

Reaction	$i+j=4n$	$i+j=4n+2$
Δ	supra-antara	supra-supra
	or antara-supra	or antara-antara
hv	supra-supra	supra-antara
	or antara-antara	or antara-supra

[1] For sigmatropic reactions involving an inversion of configuration the rules are reversed.

By applying these concepts to [1,j] sigmatropic shifts involving neutral conjugated polyenes (in an all *cis*-arrangement), general symmetry rules can be written and are summarised in table 2.4. For an [i,j] sigmatropic reaction, where i and j are both greater than unity, two π-systems are involved. The migration of the σ-bond is related to the faces of both π-systems and symmetry arguments, parallel to those above, lead to the generalised orbital symmetry rules summarised in table 2.5. However, these reactions can also be described in terms of components appropriate for the operation of the Woodward–Hoffmann generalised rule. Nevertheless, we include the rules in tables 2.4 and 2.5 since they are still found by some to be useful.

The symmetry rules (table 2.5) applied to the [3,3] reaction, the Cope rearrangement, show it is allowed in the ground state as a suprafacial–suprafacial process. This reaction can be described as a $[_\pi 2_s + _\sigma 2_s + _\pi 2_s]$ reaction. All three components are of the $(4q+2)_s$ type and the reaction is symmetry allowed from the Woodward–Hoffmann generalised rule.

Reactions of type (iii) described at the beginning of this section are crossed $[_\pi 2 + _\pi 2]$ cycloaddition reactions. This process is a specific type of cycloaddition reaction, and examples are discussed in chapter 4.

[1,3] Hydrogen shifts have been observed[22] in the singlet state photochemistry of simple alkenes. For example, irradiation of isopropylidenecyclopentane gives both of the possible products of [1,3] hydrogen shift, but it is not clear why the endocyclic alkene is preferred. Labelling studies have unambiguously established the intramolecular nature of these shifts.

Several rearrangements of 1,5-dienes have been observed and the following example (scheme 2.16) demonstrates that the reaction course depends on the electron state, S_0, S_1, T_1, from which the reaction occurs. The unsensitised (S_1) photochemical reaction leads to the two possible [1,3] sigmatropic shift products (35) and (36) while photolysis in the presence of a sensitiser results in cross-cycloaddition. Under thermal conditions the diene undergoes the Cope [3,3] sigmatropic rearrangement to give (37). This latter product is absent in the singlet excited state reaction despite the fact that a further [1,3] sigmatropic shift of C_1 to $C_{1'}$ in (35) or of $C_{1'}$ to C_1 in (36) would give this product, *viz.* the thermal [3,3] shift could be achieved by two consecutive [1,3] shifts in the excited singlet state.

Deuterium labelling experiments have demonstrated the validity of the [1,3] and [3,3] sigmatropic rearrangements as shown in scheme 2.17. The photochemical reaction leads to a product with the deuterium still attached

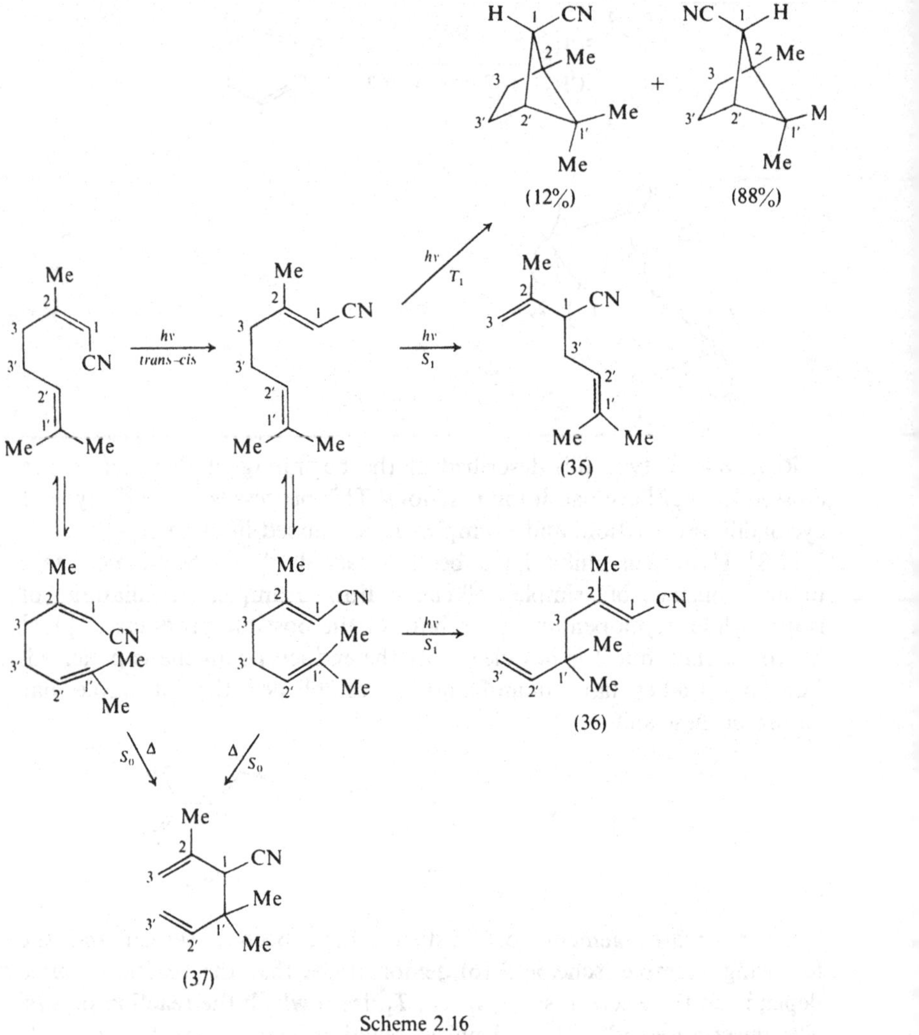

(12%)   (88%)

(35)

(36)

(37)

Scheme 2.16

to a double-bond carbon. This is a result consistent only with a [1,3] shift of C_4 from C_3 to C_1. A further [1,3] shift of C_1 to C_6 does not occur because the light absorbing dicyanoethene moiety is no longer present. Similarly photolysis of the *cis*- and *trans*-isomers of 3-methyl-5-phenyldicyano-methylenecyclohexane (38) gives rise, by a [1,3] benzylic shift, to *cis*- and *trans*-6,6-dicyano-3-methyl-5-phenylmethylenecyclohexane respectively. Each reaction is stereospecific, with retention of configuration of the migrating benzylic centre, in accord with the orbital symmetry rules for concerted photochemical [1,3] sigmatropic reactions. The photolysis of *cis*-2-methyl-

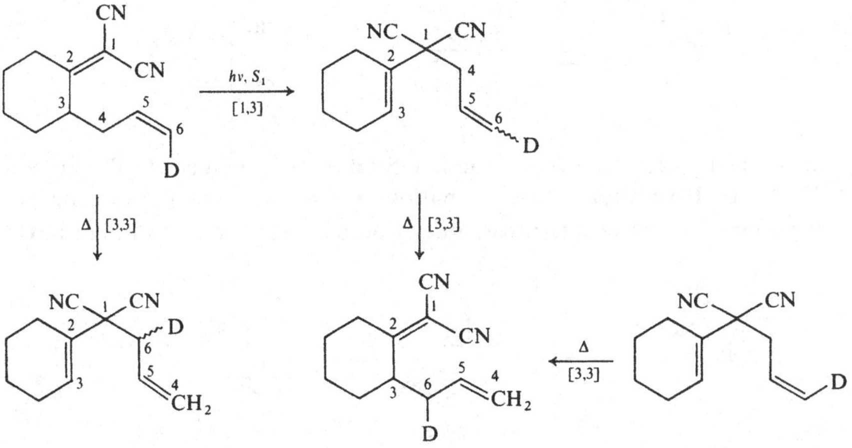

Scheme 2.17

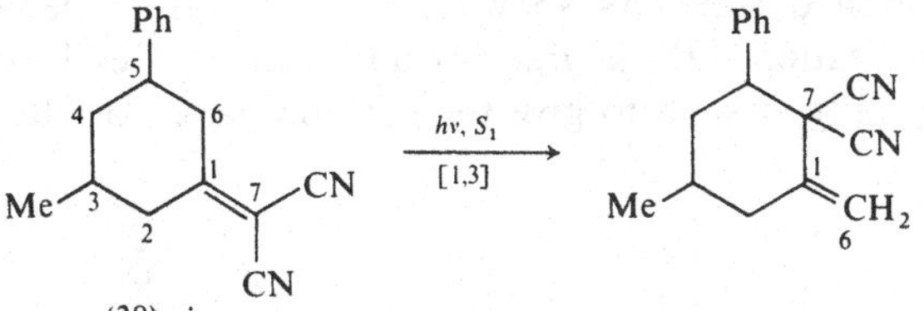

(38) *cis*

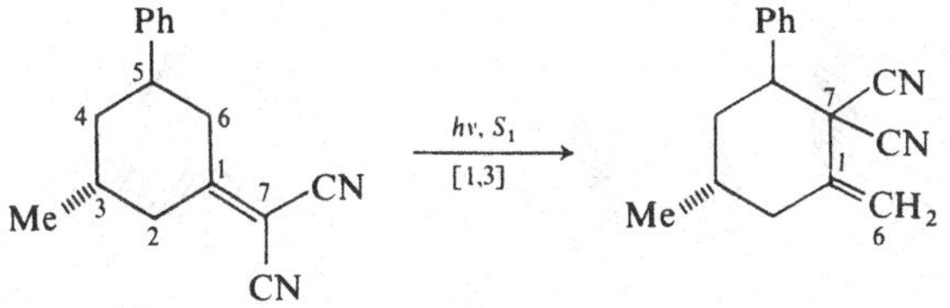

(38) *trans*

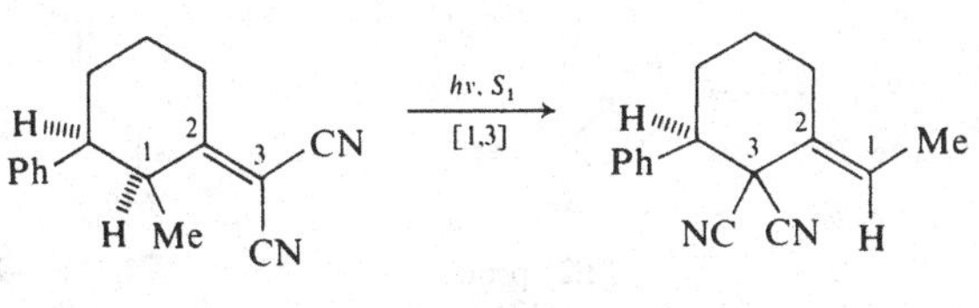

3-phenyldicyanomethylenecyclohexane also leads to a [1,3] benzylic shift. The product has the methyl group on the side of the double bond opposite to the dicyano group; this is consistent with a [1,3] benzylic shift with retention of configuration at C_1 of the dicyanoallyl group.

A different type of sigmatropic reaction has been observed for a variety of acyclic alkyl substituted 1,5-dienes.[23] The reaction occurs on direct irradiation and in competition with the [1,3] sigmatropic change discussed above. The product of this rearrangement is an allylcyclopropane and it

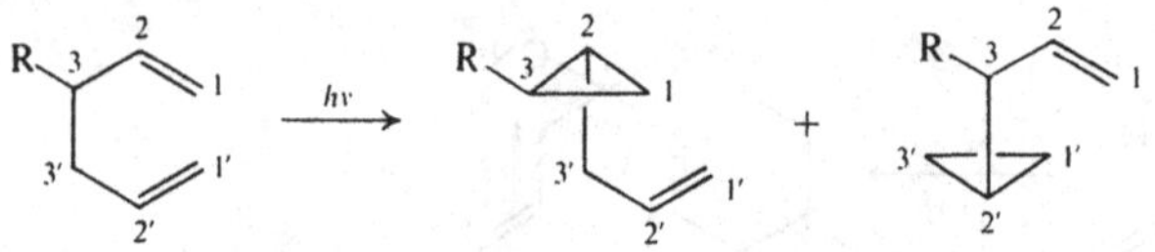

R
3
2
1
3′
1′
2′
hν
+

arises by a [1,2] sigmatropic reaction. Studies of a number of substituted 1,5-dienes have shown that formation of the three-membered ring is regioselective and occurs preferentially across the more substituted allyl

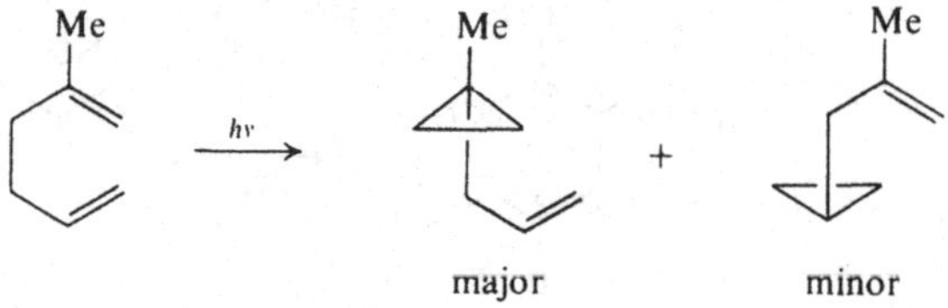

Me
hν
+
major
minor

moiety. Application of symmetry arguments to the [1,2] shift show it to be allowed in the excited state. As shown the migration proceeds with inversion of configuration at the migrating centre and requires disrotatory closure across the allyl system to give the cyclopropane. An alternative

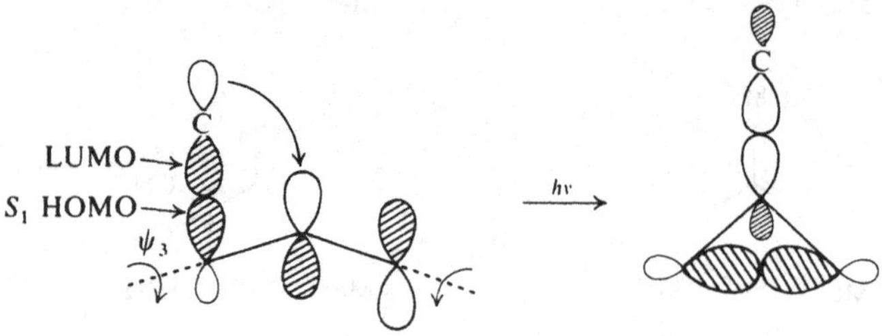

LUMO
S_1 HOMO
ψ_3
C
hν

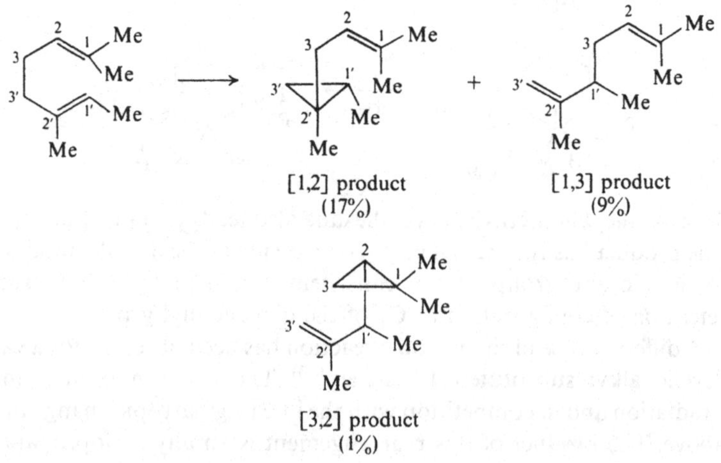

Me
+
[1,2] product
(17%)
[1,3] product
(9%)
[3,2] product
(1%)

course for the reaction would be to give product via an allowed [3,2] sigmatropic shift but such products are rarely observed. The [1,2] shift is effected when a medium pressure mercury arc is used as the light source. If direct irradiation at 254 nm, or *para*-xylene sensitisation, is employed the sole reaction observed is geometrical isomerisation of the alkene. These studies show that neither the π–π^* singlet or triplet states are involved in these allyl migrations. It has been suggested[23] that the 1,5-diene orients itself with the C_3–C_4 bond orthogonal to the plane containing the two π-bonds whereby interaction between the double bonds occurs and gives rise to a new set of molecular orbitals which can be excited.

References

1. Saltiel, J. & Charlton, J. L., in *Rearrangements in Ground and Excited States*, ed. de Mayo, P., Academic, 1980, Vol. 3, p. 25.
2. Saltiel, J., D'Agostina, J., Megarity, E. D., Mett, L., Neuberger, K. R., Wrighton, M. & Zafiriou, O. C., *Org. Photochem.*, 1973, **3**, 1.
3. Turro, N. J., *Modern Molecular Photochemistry*, Benjamin/Cummings, 1978, p. 473.
4. Dauben, W. G., van Riel, H. C. H. A., Hauw, C., Leroy, F., Joussot-Dubien, J. & Bonneau, R., *J. Am. Chem. Soc.*, 1979, **101**, 1901, and references cited.
5. Kropp, P. J., Reardon, E. J., Gaibel, Z. L. F., Williard, K. F. & Hattaway, J. H., *J. Am. Chem. Soc.*, 1973, **95**, 7058; Kropp, P. J., *ibid.*, 1969, **91**, 5783.
6. Ottolenghi, M., *Adv. in Photochem.*, 1980, **12**, 97.
7. Ross, D. L. & Blanc, J., in *Photochromism*, ed. Brown, G. H., Wiley, 1971.
8. Woodward, R. B. & Hoffmann, R., *The Conservation of Orbital Symmetry*, Verlag Chemie/Academic, 1970.
9. Gilchrist, T. L. & Storr, R. C., *Organic Reactions and Orbital Symmetry*, Cambridge Univeristy Press, 2nd edn., 1979.
10. Houk, K. N., in *Pericyclic Reactions*, eds. Marchand, A. P. & Lehr, R. E., Academic, 1977, Vol. 2, p. 182.
11. Michl, J., *Top. Curr. Chem.*, 1975, **46**, 1.
12. Fleming, I., *Frontier Orbitals and Organic Chemical Reactions*, Wiley, 1976.
13. Fukui, K., *Theory of Orientation and Stereoselection*, Springer-Verlag, 1975.
14. Dewar, M. J. S. & Dougherty, R. C., *The PMO Theory of Organic Chemistry*, Plenum, 1975.
15. Dauben, W. G., McInnis, E. L. & Michno, D. M., in *Rearrangements in Ground and Excited States*, ed. de Mayo, P., Academic, 1980, Vol. 3, p. 91.
16. Laarhoven, W. H., *Recl. Trav. Chim. Pays-Bas*, 1983, **102**, 185, 241.
17. Jacobs, H. J. C. & Havinga, E., *Adv. in Photochem.*, 1979, **11**, 305.
18. Bryce-Smith, D. & Gilbert, A., in *Rearrangements in Ground and Excited States*, ed. de Mayo, P., Academic, 1980, Vol. 3, p. 349.
19. Zimmerman, H. E., in *Rearrangements in Ground and Excited States*, ed. de Mayo, P., Academic, 1980, Vol. 3, p. 131; *Acc. Chem. Res.*, 1982, **15**, 312.
20. Morrison, H., *Acc. Chem. Res.*, 1979, **12**, 383.
21. Gajewski, J. J., *Acc. Chem. Res.*, 1980, **13**, 142.
22. Kropp, P. J., Fravel, H. G. & Fields, T. R., *J. Am. Chem. Soc.*, 1976, **98**, 840.
23. Manning, T. D. R. & Kropp, P. J., *J. Am. Chem. Soc.*, 1981, **103**, 889.

3 Intramolecular reactions of the carbonyl group

There are many carbonyl group reactions initiated by n–π* excitation and reaction can occur from both the singlet and triplet excited states. The ease and efficiency of intersystem crossing (ISC), particularly for aryl and unsaturated conjugated ketones, facilitate reactions from the triplet state and account for the ability of carbonyl compounds to act as triplet sensitisers. The lowest energy forms of the singlet and triplet excited states differ slightly in geometry as well as energy and consequently the course of a reaction may be determined by the excited state from which the reaction proceeds.

3.1 Saturated acyclic carbonyl compounds[1-8]

The photochemical reactions of saturated acyclic carbonyl compounds are dominated by three reaction processes known as *Norrish Type I*, *Norrish Type II* and *photoreduction*. A discussion of photoreduction is left to section 5.3 since the reactions are intermolecular.

The Norrish Type I process[1-3] is characterised by initial cleavage of the carbonyl–carbon bond to give an acyl and an alkyl radical. The reaction is

$$R_2CHCCR_3 \xrightarrow{h\nu} R_2CHC\cdot + R_3C\cdot$$
$$\underset{O}{\|} \qquad\qquad \underset{O}{\|}$$

initiated by n–π* excitation since aryl ketones, which have π–π* as the lowest energy triplet state, cleave only slowly if at all. Furthermore, the n–π* triplet cleaves more rapidly than the corresponding singlet despite the fact that cleavage from the singlet is more exothermic. These radical fragments can undergo stabilisation by one of the routes (*a*)–(*c*):

(*a*) α-Hydrogen abstraction by the alkyl radical to form a ketene and alkane. The presence of a ketene as a reactive intermediate has been demonstrated by spectroscopic methods and, in the presence of a nucleophilic species such as water or methanol, it can be trapped as the carboxylic acid or ester derivative.

$$\underset{\overset{|}{H}}{R_2C}-\dot{C}=O+\cdot CR_3 \longrightarrow R_2C=C=O+R_3CH$$

(*b*) Decarbonylation of the acyl radical to give carbon monoxide and an alkyl radical, the latter reacting with another alkyl radical to give an alkane

$$\underset{\overset{|}{H}}{R_2C}-\dot{C}=O \longrightarrow \underset{\overset{|}{H}}{R_2\dot{C}}\cdot+CO$$

or undergoing intermolecular hydrogen abstraction to form an alkene and an alkane.

$$2\underset{\overset{|}{H}}{R_2\dot{C}}\cdot \longrightarrow R_2CH-CHR_2$$

$$\underset{\overset{|}{H}}{R^1{}_2\dot{C}}\cdot+\underset{\overset{|}{H}}{R^2{}_2C}-\dot{C}HR \longrightarrow R^1{}_2CH_2+R^2{}_2C=CHR$$

(*c*) Intermolecular hydrogen abstraction by the acyl radical from the alkyl radical to give an aldehyde and an alkene.

$$R^1{}_2CH-\dot{C}=O+\underset{\overset{|}{H}}{R^2{}_2C}-\dot{C}HR \longrightarrow R^1{}_2CHCHO+R^2{}_2C=CHR$$

The Norrish Type II process[4,5] which competes with α-cleavage, is characterised by intramolecular hydrogen transfer from the γ-carbon atom to the carbonyl oxygen resulting in a 1,4-diradical. The 1,4-diradical can undergo fragmentation of the $\alpha\beta$-carbon–carbon bond to give a methyl ketone via the enol and an alkene (path (*a*), scheme 3.1) (the Norrish Type II reaction), intramolecular radical recombination to form a cyclo-butanol (path (*b*)), or regenerate the starting ketone (path (*c*). The facility for γ-hydrogen transfer will depend on the conformational mobility of the substrate and the lifetime of the carbonyl excited state. If the lifetime of the excited state is short the conformational distribution of the starting ketone will be important.[5] The rate constants for hydrogen abstraction by triplet ketones are such that conformational motion of alicyclic molecules can occur before the excited state undergoes reaction.

The condensed-phase photolysis of 2,2-dimethylheptan-3-one (scheme 3.2) illustrates both the Norrish Type I and Type II processes. The Type I process occurs from both the excited singlet and triplet states and the Type II process occurs predominantly from the excited singlet state. For aryl ketones ISC is so rapid ($k_{ISC} = 10^{10} \text{ s}^{-1}$) that γ-hydrogen transfer can only

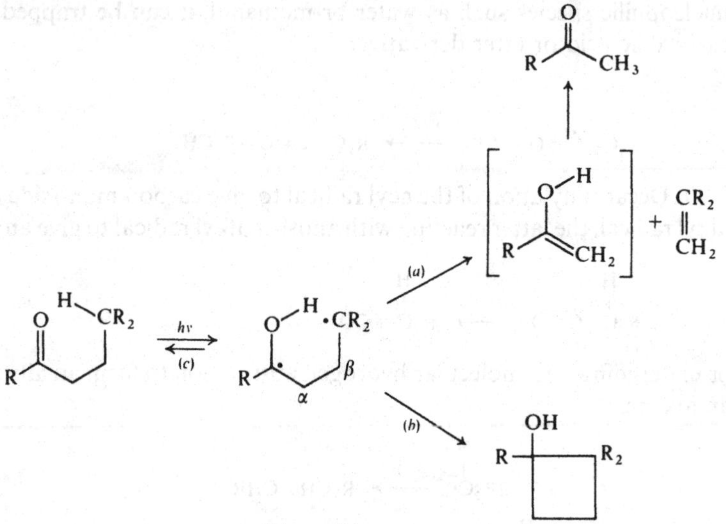

Scheme 3.1

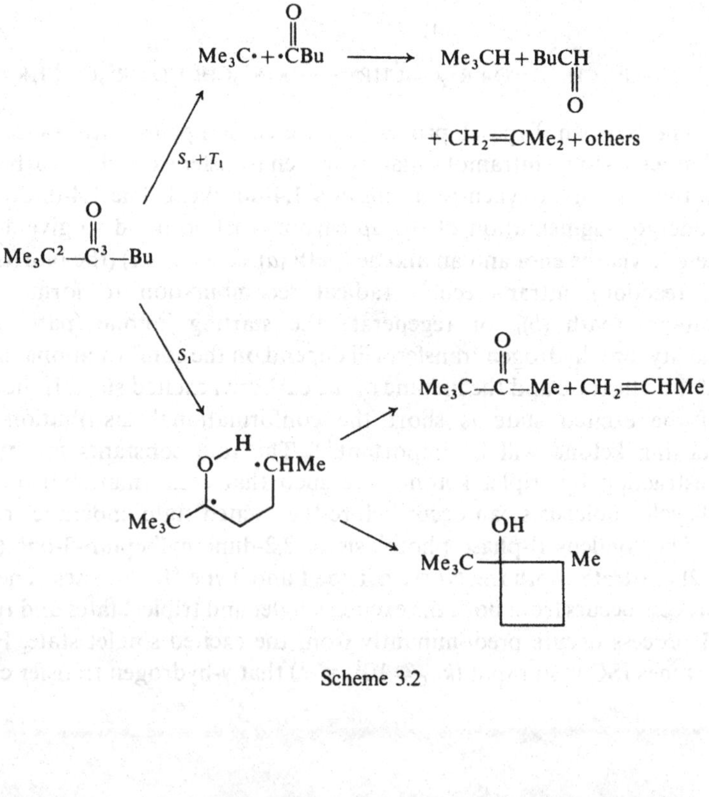

Scheme 3.2

occur in the triplet excited state. However, because of their longer singlet lifetimes ($k_{ISC} = 10^8 \, s^{-1}$) dialkylketones undergo Type II processes from both the singlet and triplet n–π* excited states.

In the Norrish Type I process there is as expected a preference to cleave the bond linking the carbonyl group to the more highly alkylated carbon. 2,2-Dimethylheptan-2-one is cleaved at the C_2–C_3 bond to give the tertiary butyl radical, which is more stable than the alternative primary butyl radical formed by cleavage of the C_3–C_4 bond. As the excitation

$$Me_3C^2 - C^3 - Bu \xrightarrow{h\nu} Me_3C^2 \cdot + \cdot C^3 - Bu \overset{O}{\underset{\|}{}}$$

$$Me_3C^2 - C^3 \cdot + \cdot Bu$$

wavelength is reduced (energy increased) the selectivity of carbon–carbonyl bond cleavage decreases. For butan-2-one little selectivity of bond cleavage is observed at 254 nm, where $\Phi_a/\Phi_b = 2.4$, but as the energy of the radiation is reduced a greater preference for cleavage of the weaker bond is observed ($\Phi_a/\Phi_b = 5.5$ at 265 nm and $\Phi_a/\Phi_b = 40$ at 313 nm).

$$Me - C - Et \overset{\Phi_a}{\underset{\Phi_b}{\rightrightarrows}} \begin{matrix} Me - \overset{O}{\underset{\|}{C}} \cdot + \cdot Et \\ Et - \overset{O}{\underset{\|}{C}} \cdot + \cdot Me \end{matrix}$$

The Type I α-cleavage can be followed by decarbonylation to give carbon monoxide and an alkyl or aryl radical. The presence of such radical intermediates is readily demonstrated by photolysis of a mixture of ketones which give products from mixed radical recombination (scheme 3.3).

Photolysis of 2,2,4,4-tetramethylpentan-3-one results in a high yield (>90%) of carbon monoxide from both the excited singlet and triplet states.

$$Me_3C - \underset{\underset{O}{\|}}{C} - CMe_3 \xrightarrow{h\nu} Me_3CH + Me_2C{=}CH_2 + Me_3C - CMe_3 + CO$$

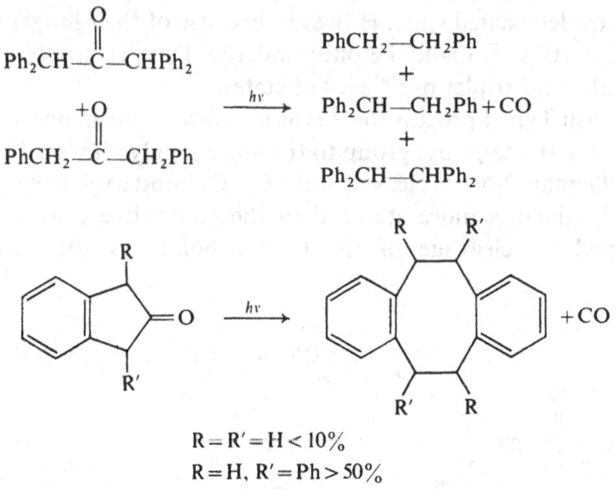

$$R = R' = H < 10\%$$
$$R = H, \ R' = Ph > 50\%$$

Scheme 3.3

The lifetime of the singlet state is $(4.5-5.6) \times 10^{-9}$ s compared with 0.11×10^{-9} s for the excited triplet state. Since the reaction occurs from both the singlet and triplet excited states, the Type I process must occur approximately 100 times faster from the triplet than from the singlet excited state.

The Type I process is favoured by photolysis in the vapour phase and is less pronounced for photolysis in inert solvents, where the solvent cage facilitates recombination of the initially generated radical pair. This is reflected in a lower quantum yield of products formed from the Type I process in inert solvents. When the Type I process does occur for an unsymmetrical ketone it is generally observed that cleavage of the more substituted α-bond occurs to give the more stable radicals. For example, photolysis of 2,2-dimethylheptan-3-one in solution is dominated by α-cleavage, which reflects the stability of the t-butyl and acyl radicals produced. Studies with triplet quenchers, such as penta-1,3-diene, have shown that in this reaction the Type I process occurs from both the singlet and triplet excited states and the Type II process occurs from the singlet state.† This result contrasts with the unsensitised photolysis of pentan-2-one in which the Type II reaction proceeds from both excited states. The efficiency of ISC varies with wavelength and therefore the importance of

---

† Care must be exercised in making an accurate measurement of the relative singlet and triplet quantum yields from quenching data; fluorescence of pentan-2-one, 2,2-dimethylbutan-3-one and acetone is quenched at high penta-1,3-diene concentrations, indicating that the singlet state of the excited carbonyl is affected by the diene.

each excited state is dependent on the wavelength of light used. Polar solvents enhance the yield of Type II photoreaction, presumably by stabilising the triplet excited state. Reactions involving the excited singlet state are less sensitive to solvent polarity.

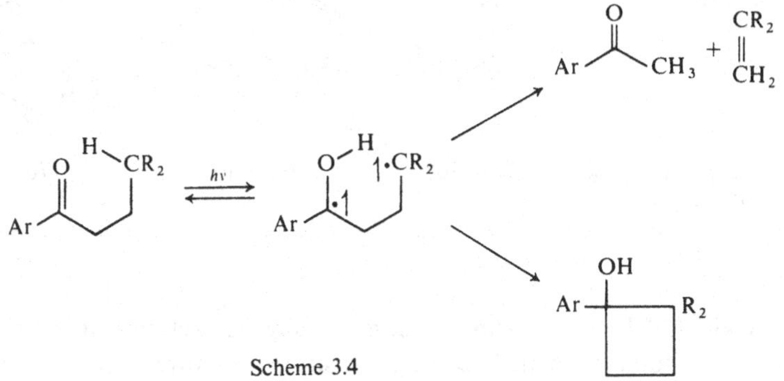

Scheme 3.4

For aryl alkyl ketones (scheme 3.4), both the reaction to give cyclobutanols, and the cleavage to form an alkene and an enol (which isomerises to a methyl ketone) are thought to proceed by way of a triplet state reaction and a 1,4-diradical.[6] The diradical may, of course, revert to starting material by reabstraction of hydrogen by the γ-carbon radical thereby reducing the quantum efficiency for the formation of cyclobutanol and fragmentation products. The intermediacy of 1,4-diradicals in these reactions has been established by intercepting them chemically with scavengers (HBr, RSH, O_2, R_2C=Se), and by direct spectroscopic observations; simple 1,4-diradicals have lifetimes of the order of 10^{-7}– 10^{-8} s. The hydrogen transfer to the oxygen atom has been shown to be intramolecular with a six-membered cyclic transition structure since 5,5-dideuterohexan-2-one gives 1-deuteroacetone and 2-deuteropropene (scheme 3.5). γ-Hydrogen abstraction is some twenty times faster[4] than

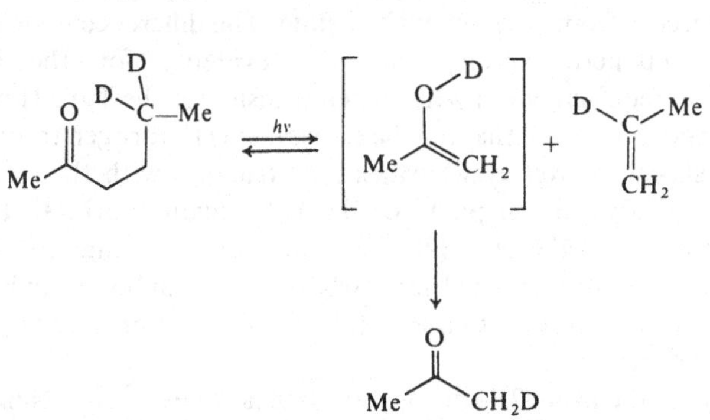

Scheme 3.5

Table 3.1 *Rate constants for γ-hydrogen transfer*

Structure			
Rate constant, k_{trf} (s^{-1})	1.2×10^8	6.0×10^8	70×10^8

equivalent δ-hydrogen abstraction in acyclic ketones indicating the preference for a six-membered ring transition structure. Any rate constants for a γ-hydrogen abstraction will include the inherent rate of hydrogen transfer and a conformational factor reflecting the proportion of the various possible conformers. For molecules where the preferred ground state conformation is such that the carbonyl is proximate to a γ-hydrogen the rate of hydrogen transfer is increased as illustrated by reaction of the substrates shown in table 3.1.

The transfer of hydrogen in the excited state occurs to the localised, half-filled, n-orbital of the oxygen atom which lies in the nodal plane of the π-bond. The *cis*-dialkyl ketone (1), with an equatorial alkyl group at C_2 undergoes a Type II reaction to give 4-*t*-butylcyclohexanone (scheme 3.6). The reaction is unaffected by the addition of penta-1,3-diene, which implies a singlet excited state process. On the other hand, the *trans*-dialkylketone (2) (scheme 3.6) undergoes isomerisation by Type I α-cleavage of the C_1–C_2 ring bond and subsequent radical recombination to the more stable *cis*-isomer. This reaction is markedly quenched by penta-1,3-diene and therefore proceeds from a triplet excited state. The difference in behaviour of the two epimeric ketones provides evidence for the specific stereochemical requirement in γ-hydrogen transfer for the Type II process. Molecular models indicate that for the *cis*-isomer (1) hydrogen transfer can occur by a six-membered cyclic transition structure with the γ-carbon–hydrogen bond axis in the plane of the half-vacant n-orbital. For the isomeric axial propylketone (2), inversion of the ring to a boat conformation is required before the γ-carbon–hydrogen bond can lie in the plane of the carbon–oxygen double bond and the reaction is consequently much less favoured.

Irradiation of the *trans*-2,6-dipropylcyclohexanone (3)[2] results in loss of the equatorial propyl side chain, rather than the axial alternative, and yields

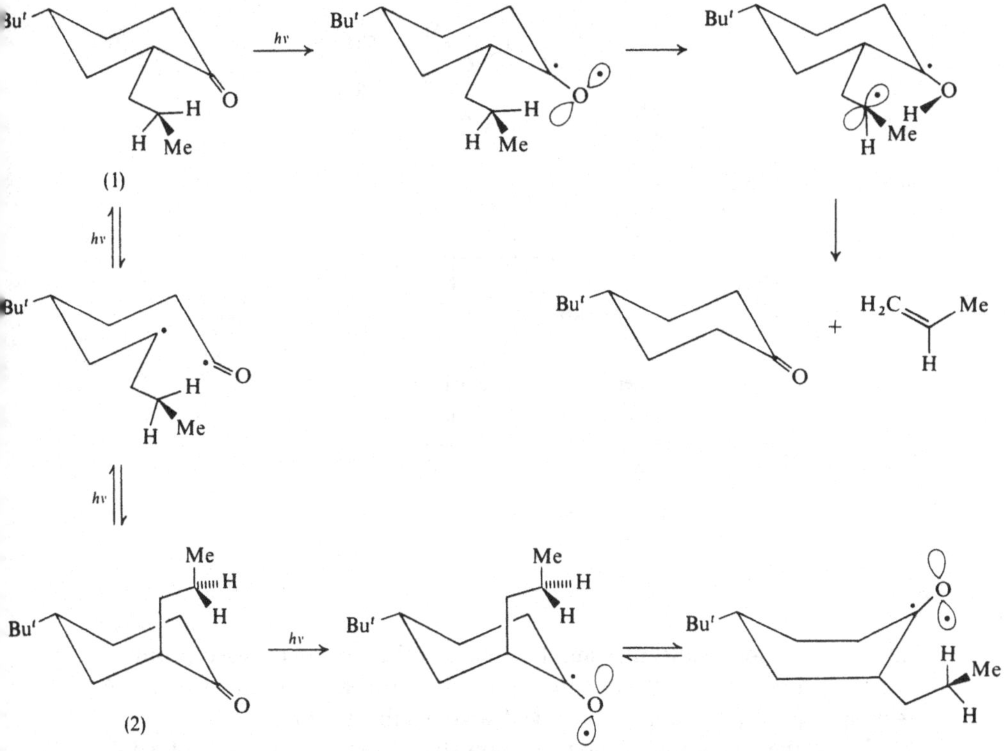

Scheme 3.6

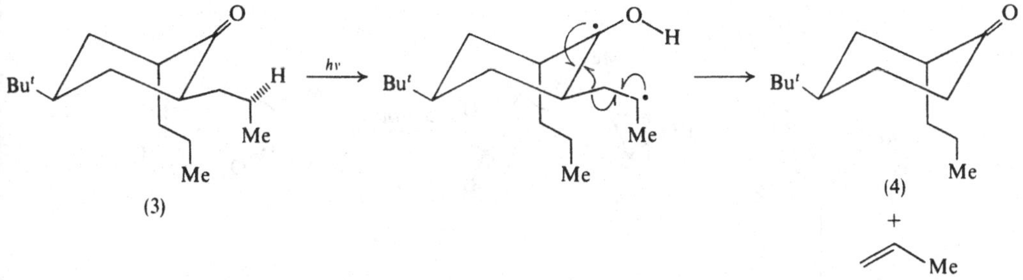

the less thermodynamically stable axial ketone (4) which is relatively stable to further irradiation. This again illustrates the more favourable six-membered ring transition structure for γ-hydrogen transfer.

The partitioning of the 1,4-diradical (formed by γ-hydrogen abstraction) between fragmentation and ring closure is dependent upon the facility of a given diradical to undergo cyclisation or fragmentation. This is influenced by α-substitution (scheme 3.7). The presence of a single methyl substituent

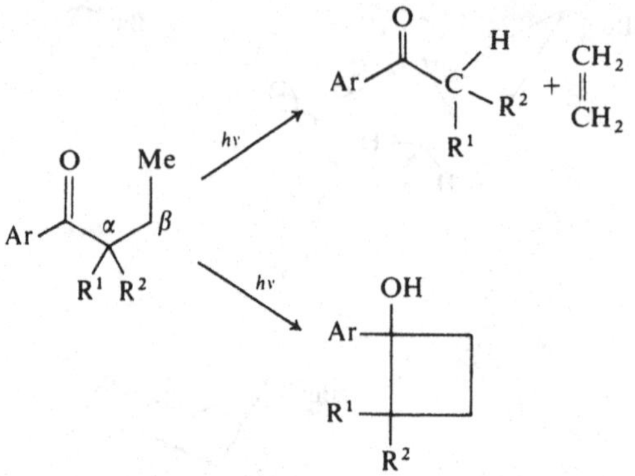

Substituents	Cyclisation
$R^1 = R^2 = H$	10%
$R^1 = H$; $R^2 = Me$	29%
$R^1 = R^2 = Me$	89%

Scheme 3.7

at the α-position results in a small increase in the yield of cyclised product, whereas α,α-dimethyl disubstitution results in a substantial increase. A similar, but minor, effect is observed with γ-substitution.

Whilst the formation of the transition structure for cyclisation of a 1,4-

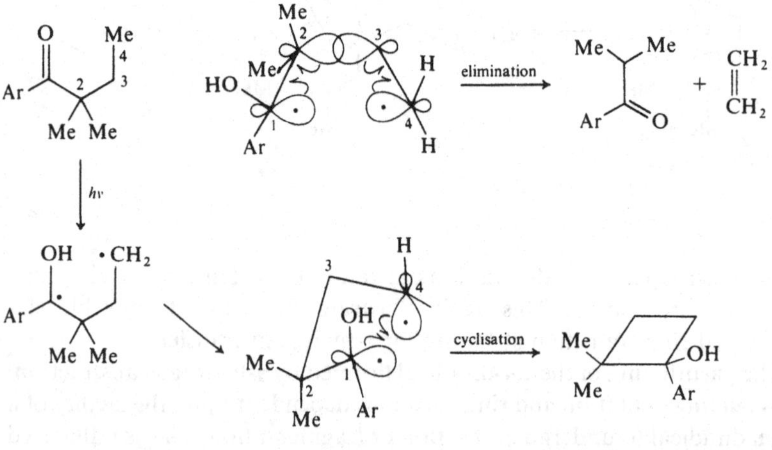

Scheme 3.8

diradical intermediate only requires overlap of the radical centres, carbon–carbon bond cleavage requires the radical centres to overlap with the bond undergoing cleavage. This is shown in scheme 3.8 for one possible conformation where this requirement is met. The magnitude of the resultant eclipsing interactions, primarily along the C_1–C_2 bond, will be greatest for elimination. With substituents at C_2, adjacent to the carbonyl group, an increase in the magnitude of these eclipsing interactions will favour cyclisation by way of the non-planar, 1,4-diradical. With β-substituents the eclipsing energy along the C_1–C_2, C_2–C_3 and C_3–C_4 bonds for cyclisation is not increased significantly, while 1,3-diaxial interactions, particularly for β,β-dimethyl substitution, will favour elimination (scheme 3.9).

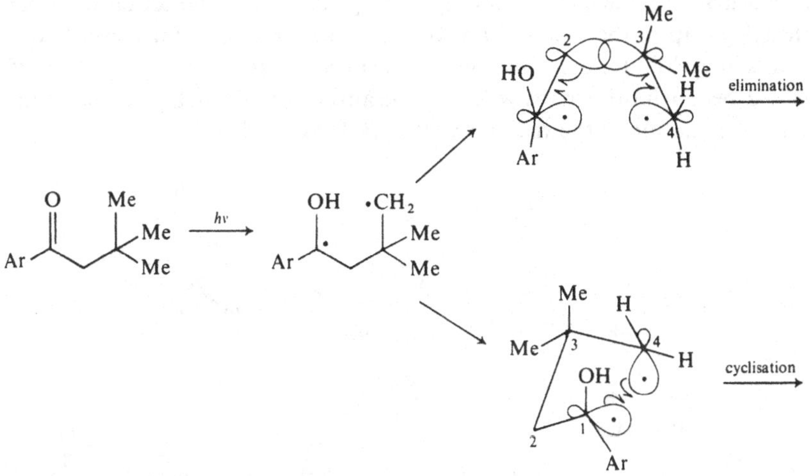

Scheme 3.9

The adamantyl ketone (5) is an example of a ketone in which γ-hydrogen abstraction results exclusively in cyclobutanol formation. The alternative elimination reaction path would produce a highly strained adamantene and would require overlap of the radical orbitals with the bond undergoing cleavage. This cannot be obtained for the 1,4-diradical because of the conformational restraints of the adamantane structure. In a similar manner *endo*-2-benzoyl-2-methylnorbornane results in quantitative formation of the

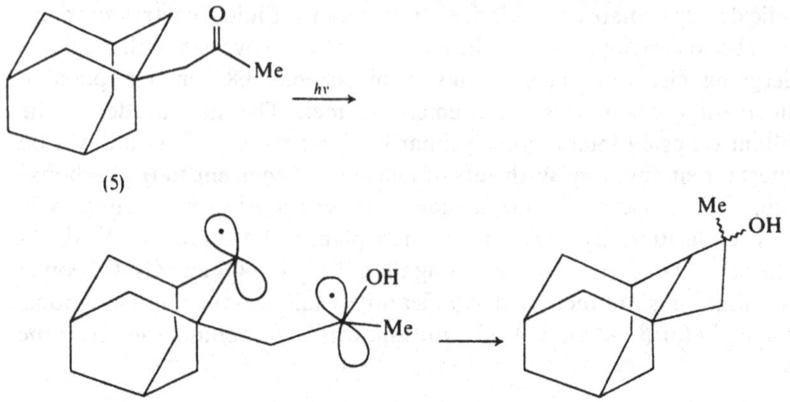

(5)

tricyclooctane (scheme 3.10, R = Me). The presence of the 2-methyl substituent, and the consequent high-energy eclipsing interaction with the phenyl group in the planar transition structure required for elimination, results in cyclobutanol formation. This is in contrast with the photolysis of *endo*-2-benzoylnorbornane where, in the absence of the methyl substituent, photoelimination is the major reaction (scheme 3.10, R = H).

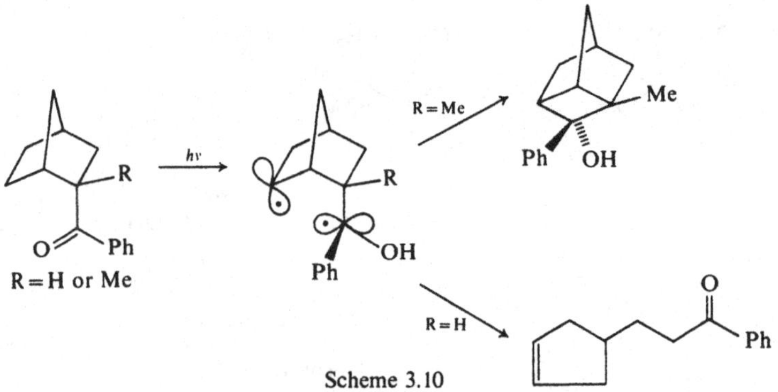

Scheme 3.10

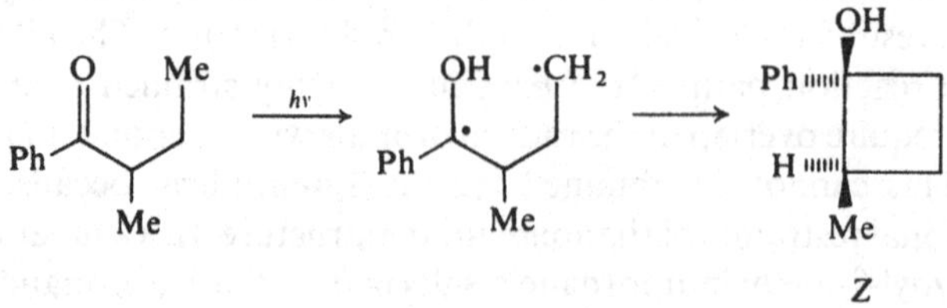

Often for aryl alkyl ketones more than one stereoisomeric cyclobutanol can be formed. The sole formation of the (Z)-cyclobutanol from 1-phenyl-2-methylbutan-1-one reflects repulsion between the methyl and phenyl groups in the 1,4-diradical. In contrast both isomeric cyclobutanols are formed from 1-phenylpentan-1-one. For this latter substrate the 1,4-methyl–phenyl interaction is small until the 1,4-bond is almost completely formed.

Photolysis of the epoxyketone (6) gives a high yield of a 1.7:1 mixture of epoxycyclobutanols (8) and (9).[7] The 1,4-diradical species is formed in a conformation where the radical orbitals are orthogonal[7]. The higher yield of

stereoisomer (8) indicates a preference for clockwise rotation about the C_1–C_2 bond of this diradical species. This reflects greater steric compression between the epoxide oxygen and the phenyl group than between the epoxide and hydroxyl group. The steric interaction will be small until the 1,4-bond is well developed and the groups are close together. In the photolysis of the methyl substituted epoxy–ketones (10) and (11), the steric interaction between the methyl group at C_2 and the substituent at C_1 determines the direction of rotation about the C_1–C_2 bond and thus the stereochemistry of cyclisation (scheme 3.11).

The structural constraint of the three-membered ring upon the diradicals of scheme 3.11 compels overlap of the radical orbital on the terminal carbon of the epoxide and the C_2–C_3 bond; this constraint would be expected to favour fragmentation. The absence of more than traces of fragmentation

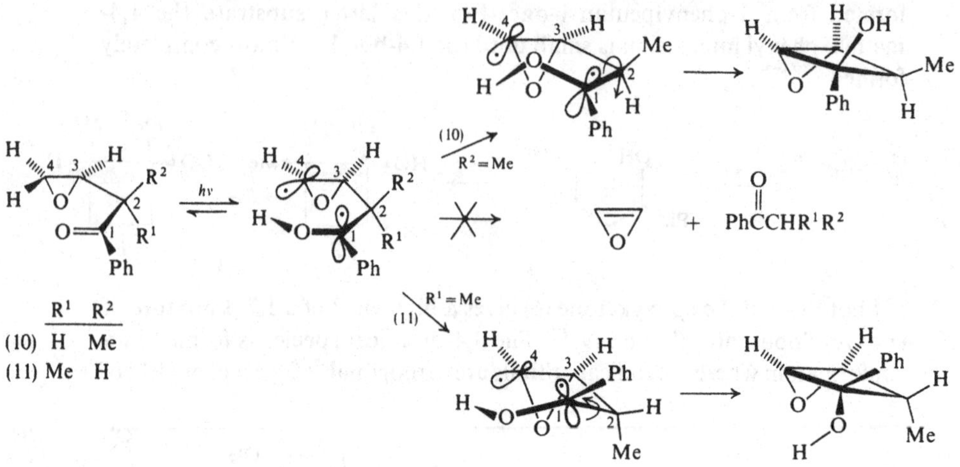

Scheme 3.11

products from these epoxides suggests that the high energy of oxirene, the expected primary fragmentation product from both epoxides (scheme 3.11), provides to high an energy barrier. Despite the presence of the C_2-methyl group in these epoxides α-cleavage is not competitive. However, the C_2-dimethyl analogue (12) exhibits α-cleavage as the dominant process.

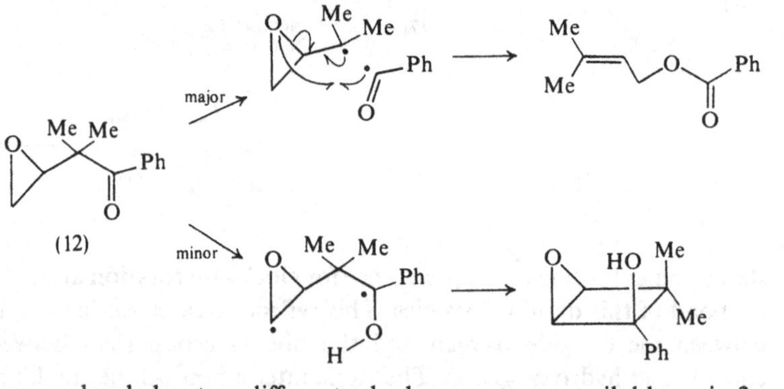

Where a molecule has two different γ-hydrogen atoms available, as in 3-methylpentanal (13), transfer in the Type II process is marked by a preference for cleavage of the weaker secondary carbon–hydrogen bond.

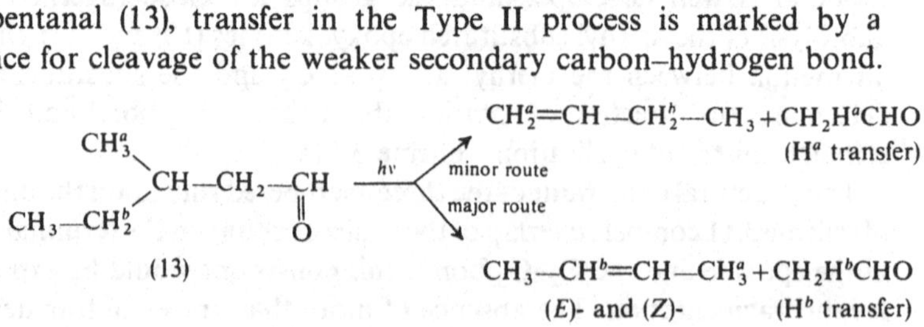

An increase in temperature or in the energy of the absorbed radiation, results in less discrimination and an increase in formation of the thermodynamically less stable (Z)-but-2-ene.

For alkyl aryl ketones substituents are important in influencing the course and efficiency of reaction. Electron releasing *para*-methyl and *para*-methoxy substituents decrease the rate constant and the quantum yield for Type II cleavage. Following this trend *para*-hydroxy, *para*-amino and

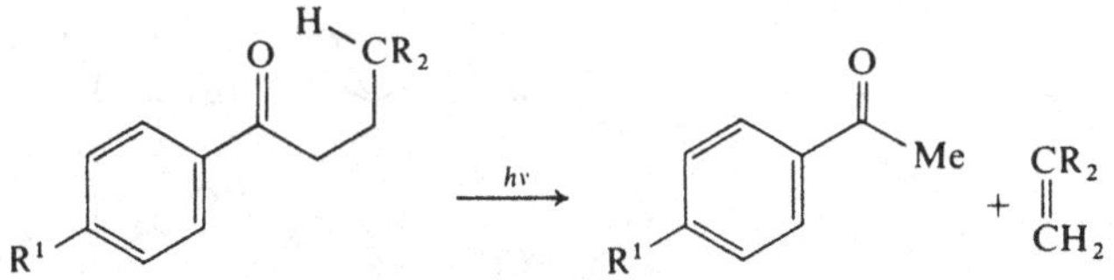

para-phenyl substituents inhibit the reaction completely. This effect implies an increase in importance of the $\pi-\pi^*$ triplet as it becomes lower in energy than the $n-\pi^*$ triplet state. If there is no γ-hydrogen atom available abnormal transition structures may be favoured as shown by the two examples of scheme 3.12. A spectacular example[8] of a hydrogen

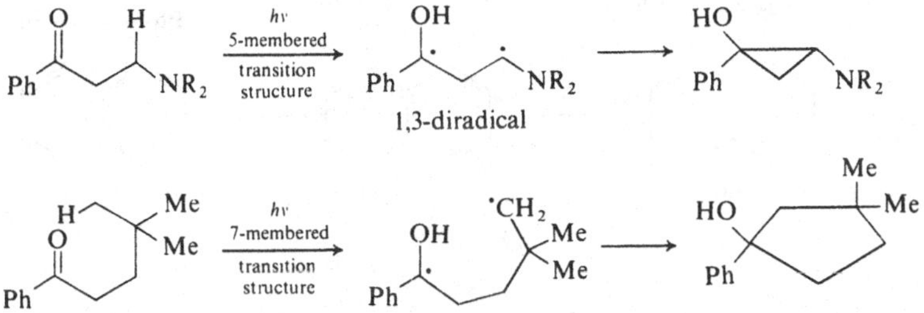

Scheme 3.12

abstraction at a remote site by an excited carbonyl is shown in scheme 3.13 and demonstrates something of the imaginative use to which the radical nature of an excited carbon can be put.

The naphthyl ketone (14), in the absence of a γ-hydrogen atom, undergoes hydrogen transfer from the δ-position via a seven-membered transition structure. The T_1 ($\pi-\pi^*$) excited state of the naphthyl ketone is lower in energy ($240\ kJ\ mol^{-1}$) than the T_2 ($n-\pi^*$) excited state ($180\ kJ\ mol^{-1}$) and the activation energy for the reaction is $c.\ 21\ kJ\ mol^{-1}$. It has been suggested that the $\pi-\pi^*$ excited state rises in energy as the reaction proceeds and, at some intermediate geometry between ketone and diradical, a surface crossing between the excited states occurs.[3]

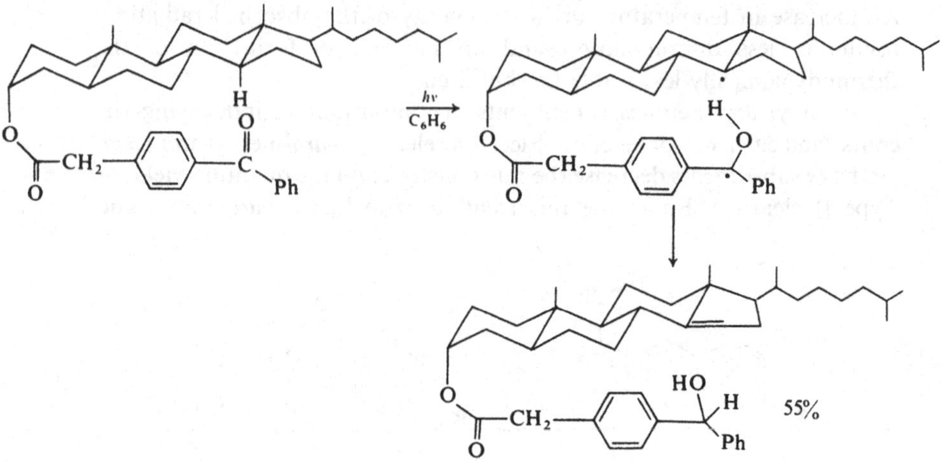

Scheme 3.13

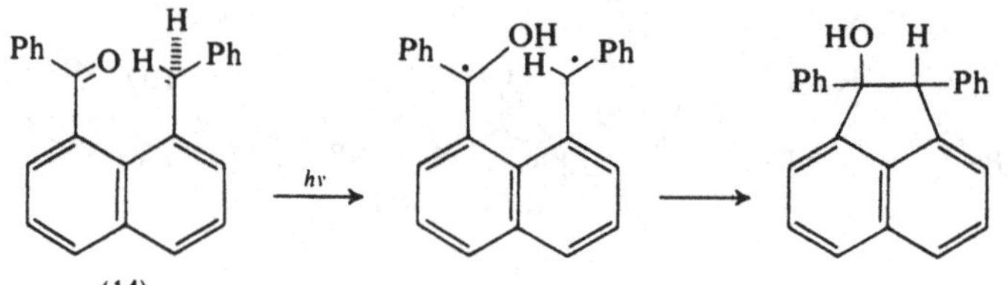

(14)

3.2   Saturated cyclic carbonyl compounds[1,2,4,9–12]

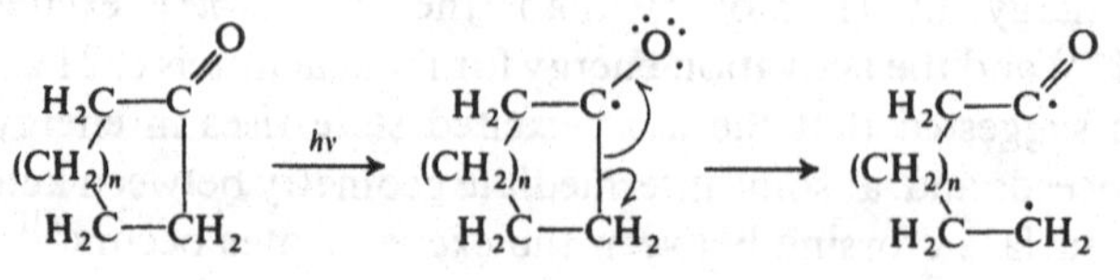

The photochemistry of saturated cyclic carbonyl compounds is dominated by the Norrish Type I process involving the initial cleavage of a carbon–carbonyl bond. The reaction requires $n-\pi^*$ excitation since those ketones in which the $\pi-\pi^*$ triplet state is of lower energy either cleave slowly or not at all. While the $n-\pi^*$ singlet state is generally less reactive to α-cleavage than is the $n-\pi^*$ triplet, many α-cleavage reactions proceed from both excited states.[2] Subsequent processes, (a)–(c), correspond to those observed for acyclic carbonyl compounds:

(a) Intramolecular α-hydrogen abstraction by the terminal alkyl radical to produce a ketene. In the case where $n = 0$, ketene and ethlene are formed.

$$
\begin{array}{c}
\text{CH}{-}\overset{\cdot}{\text{C}}\overset{\displaystyle O}{{=}} \\
(\text{CH}_2)_n \quad \text{H} \\
\text{CH}_2{-}\overset{\cdot}{\text{CH}}_2
\end{array}
\quad\longrightarrow\quad
\begin{array}{c}
\text{CH}{=}\text{C}{=}\text{O} \\
(\text{CH}_2)_n \\
\text{CH}_2{-}\text{CH}_3
\end{array}
$$

$$
\begin{array}{c}
\text{CH}_2{-}\overset{\displaystyle O}{\overset{\|}{\text{C}}}\cdot \\
| \\
\text{CH}_2{-}\overset{\cdot}{\text{CH}}_2
\end{array}
\quad\longrightarrow\quad
\begin{array}{c}
\text{CH}_2{=}\text{C}{=}\text{O} \\
+ \\
\text{CH}_2{=}\text{CH}_2
\end{array}
$$

(b) Intramolecular hydrogen abstraction by the carbonyl carbon atom to give an unsaturated aldehyde.

$$
\begin{array}{c}
\text{CH}_2{-}\overset{\displaystyle O}{\overset{\|}{\text{C}}}\cdot \\
(\text{CH}_2)_n \quad \text{H} \\
\text{CH}_2{-}\overset{\cdot}{\text{CH}}_2
\end{array}
\quad\longrightarrow\quad
\begin{array}{c}
\text{CH}_2{-}\text{CHO} \\
(\text{CH}_2)_n \\
\text{CH}{=}\text{CH}_2
\end{array}
$$

(c) Photo-decarbonylation to give carbon monoxide, an alkene or a cyclic alkane or both of the latter.

$$
\begin{array}{c}
\text{CH}_2{-}\overset{\displaystyle O}{\overset{\|}{\text{C}}}\cdot \\
(\text{CH}_2)_n \\
\text{CH}_2{-}\overset{\cdot}{\text{CH}}_2
\end{array}
\quad\longrightarrow\quad
\begin{array}{c}
\text{CH}_2\cdot \\
(\text{CH}_2)_n \qquad +\,\text{CO} \\
\text{CH}_2{-}\overset{\cdot}{\text{CH}}_2
\end{array}
$$

radical recombination

$$
\begin{array}{c}
\text{CH}_2 \\
(\text{CH}_2)_n \quad \text{CH}_2 \\
\text{CH}_2
\end{array}
$$

intramolecular hydrogen abstraction

$$
\begin{array}{c}
\text{CH}_3 \\
(\text{CH}_2)_n \\
\text{CH}{=}\text{CH}_2
\end{array}
$$

The major competitive triplet state reaction for cycloalkanones which are not substituted at the α-carbon is photoreduction by solvent which proceeds with a pseudo-unimolecular rate constant of approximately $10^6 \, s^{-1}$ in hexane and $10^7 \, s^{-1}$ in ethanol. In t-butyl alcohol, acetone, acetonitrile, acetic acid and ethyl acetate photoreduction is much slower $(k = 10^4 \, s^{-1})$ and therefore less competitive (see section 5.3).

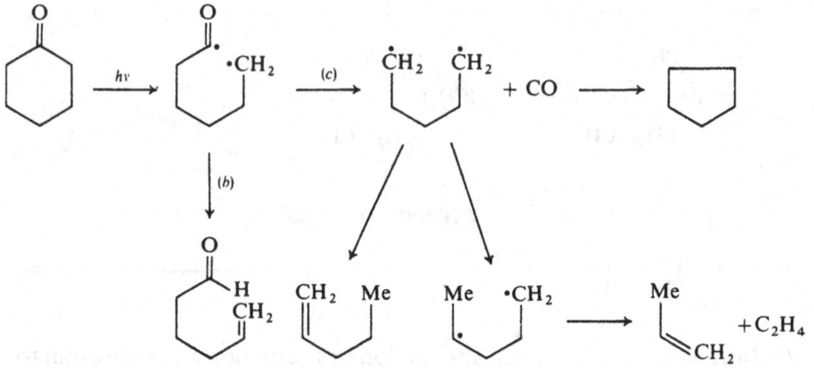

Scheme 3.14

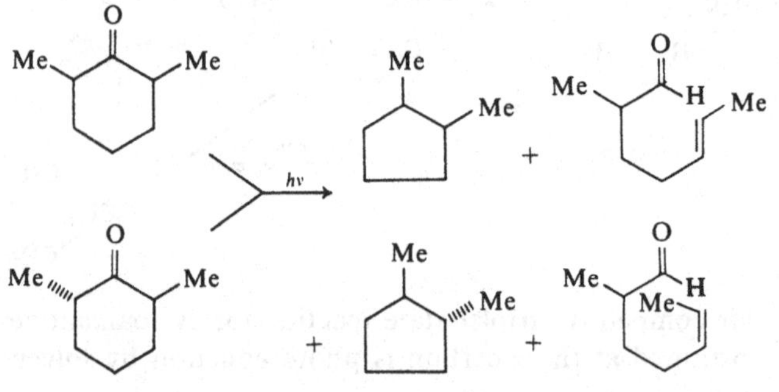

The vapour-phase photolysis of cyclohexanone (scheme 3.14) affords products from both pathways (b) and (c). The 1,6-triplet diradical from Norrish Type α-cleavage of cyclohexanone has been detected[12] and ISC from triplet to singlet is the lifetime determining process of the diradical. The lifetimes of such diradicals are in the range of twisted alkene triplets and 1,4-diradicals. The intermediacy of a diradical in these reactions also follows from the results of vapour-phase photolysis of the epimeric 2,6-dimethylcyclohexanones. In each case, product formation is faster than photochemical interconversion of the epimers and each epimer gives the same mixture of products. This shows that the configurational integrity of the α-carbon is lost during the course of the reaction; this is consistent with free rotation in an initially produced diradical. If there is a proximate γ-hydrogen the excited carbonyl can undergo γ-hydrogen transfer (Norrish Type II reaction) even in a cyclic system. Moreover, γ-hydrogen transfer in the singlet state is rapid and thus this reaction can occur prior to ISC to the triplet from which α-cleavage would be expected.

The photochemical epimerisation of 17-ketosteroids is an example of an α-cleavage reaction where reclosure of the acyl-alkyl diradicals occurs to give the thermodynamically more stable 17-keto compound with the *cis*

C/D ring junction. These epimerisation reactions have low quantum yields which suggest that the process is reversible. Each of the *cis*- and *trans*-fused bicyclic ketones (15*a*) and (15*b*) undergo cage recombination of the

(15*a*) $n = 1$
(15*b*) $n = 2$

diradicals formed by Type I cleavage, and there is consequent isomerisation at the ring junction. This isomerisation is not affected by the addition of triplet quenchers and the reaction is therefore regarded as involving the singlet excited state. Aldehyde formation, which corresponds to pathway (*b*) in scheme 3.14, occurs from both excited states.

The photolysis of cyclopentanone and 2-methylcyclopentanone, involves Norrish Type I α-cleavage and in each case this proceeds from the singlet and the triplet excited states in the condensed phase. The photochemistry of 2,2,6,6-tetramethylcyclohexanone has been examined and the excited states S_1 and T_1 are again involved. This is in contrast with the conclusions drawn from studies on cyclohexanone, 2- and 3-methylcyclohexanone, and 2,2-, 2,5- and 2,6-dimethylcyclohexanone: the formation of unsaturated aldehyde by Type I cleavage and hydrogen transfer is completely quenched by dienes in benzene and therefore occurs from the triplet excited state. It has been assumed that the reaction involves exclusively the triplet state. It is possible that ISC (S_1–T_1) in this group of compounds is too rapid for reaction from the singlet state to be observed. A detailed study of 2,2,5,5-tetramethylcyclopentanone and the bicyclic ketones (16) has shown that, while these reactions were not quenched,

(16a) $R^1 = R^2 = R^3 = H$ or Me

(16b) $R^1 = R^2 = H$, $R^3 = Me$

alkenal formation did in fact occur, in part, from the triplet state. The triplet was found to be at least 100 times more reactive than the excited singlet state and comparable conclusions have also been found for other ketones.

Photolysis of 2-methylcyclohexanone (17) gives a 3:1 mixture of

trans- and *cis*-hept-5-enal with only a trace of 2-methylhex-5-enal (18). The preference for formation of the *trans*-alkenal can be rationalised by a conformational preference in the six-membered, hydrogen-bridged, cyclic transition structure leading to this product. When the pressure is reduced the vibrational energy of the system is increased (since collisional deactivation is reduced) and the selectivity of hydrogen transfer leading to the *trans*-alkenal becomes less marked. The preference for cleavage of the C_1-C_2 bond compared with the C_1-C_6 bond is at least fifty fold and is consistent with the relative stabilities of secondary and primary radicals. The vapour-phase photolysis of (17) gives a higher yield of decarbonylated product than does photolysis in the liquid phase. Moreover, the proportion increases as the pressure is reduced since the excited species are involved in fewer collisions; excitation energy is not so readily lost and decarbonylation is greatly favoured. Certain types of α-substitution force cleavage of the less substituted α-bond as exemplified by the spiro ketone (19). Cyclopropyl groups can also influence selection for cleavage

particularly when a cyclopropyl stabilised radical can be formed. For example, 4-carvanone (20) suffers α-cleavage towards the least substituted centre to generate such a stabilised species.

In protic solvents, cyclic ketones can undergo the Norrish Type I process to yield ketenes which then undergo addition of solvent to give carboxylic acids or their derivatives. This process is appropriately demonstrated by the photochemistry of carvone camphor (21), in which the diradical formed by carbonyl–carbon bond cleavage is transformed into a ketene by

intramolecular hydrogen transfer from the α-position. Deuterium labelling experiments have shown the transfer to be stereospecific and to involve only the *exo*-hydrogen. This is compatible with either *exo*-hydrogen transfer during rotation of the carbonyl group away from the two methyl substituents or the fewer non-bonded repulsions involved in *exo*- versus *endo*-hydrogen transfer, or both. For simple cyclohexanones the equivalent diradical abstracts the hydrogen atom which was initially axial twice as

exo-hydrogen   *endo*-hydrogen
transfer   transfer

rapidly as the hydrogen which was equatorial. A pseudochair transition structure for this process requires the methylene group to be pseudoaxial for transfer of the equatorial hydrogen whereas it is pseudoequatorial in the favoured process.

For carbonyl groups contained in medium-sized rings, intramolecular hydrogen transfer reactions successfully compete with α-cleavage and lead to bicyclic products. In the formation of *trans*-9-hydroxydecalin from cyclodecanone (22), a concerted transannular hydrogen transfer with ring

(22)   3:1

closure is geometrically highly unfavourable; consequently a diradical intermediate with its more favourable geometry is more likely. Cyclodecanone (22) is unique among cyclic ketones in forming only alcohol products and this must reflect the conformational preference of the ring with the carbonyl n-orbital proximate to an ε-hydrogen atom. In larger ring systems, γ-hydrogen abstraction is the dominant cyclisation mechanism.

The photochemistry of cyclobutanone[10,11] is dominated by α-cleavage. In contrast to other cycloalkanones, cyclobutanone is efficiently α-cleaved from the singlet n–π* excited state and a singlet diradical is produced.

Subsequent decarbonylation produces a 1,3-diradical which in turn gives cyclopropane, whilst cycloreversion gives ketene and ethene. The singlet diradical can also cyclise through oxygen and give a ring-expanded oxacarbene which is trapped by alcohols. In some cases an almost quantitative yield of the ring-expanded product has been reported.

When photolysed at $-78°C$ in the presence of butadiene, cyclobutanone gives 3-vinylcyclohexanone and an oxetane; the oxetane results from a $[_\pi2 + _\pi2]$ intermolecular cycloaddition (section 4.2). The use of deuterium labelled butadiene has demonstrated that the 3-vinylcyclohexanone is formed by addition of the initially formed 1,4-diradical to one double bond of the diene. The result also shows that the acyl radical is more reactive than the alkyl radical.

Irradiation of *cis*- (or *trans*-) substituted cyclobutanones in methanol leads to three characteristic solution photoreactions: decarbonylation, $[_\pi2 + _\pi2]$ cycloreversion and ring expansion (scheme 3.15). The first two are stereospecific and the latter highly stereoselective. The reactions occur from α-cleavage in the n–π* singlet excited state. In contrast other

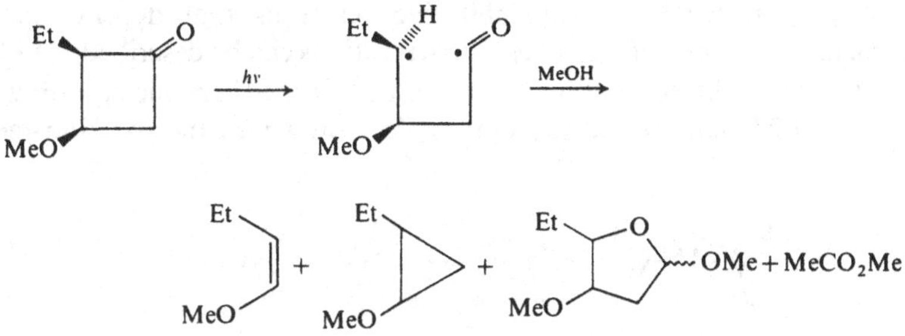

Scheme 3.15

cycloalkanones are α-cleaved from their n–π* triplet excited state. These results also contrast with the photochemical behaviour of cyclohexanones and cyclopentanones, where the intermediates formed are capable of losing stereochemical integrity, and with the vapour-phase photolysis of cyclobutanone, where carbon monoxide is lost by a triplet mechanism. The stereospecificity observed for photolysis of substituted cyclobutanones is general for singlet *cis*-1,4-carbon diradicals. In these cases subsequent stabilisation proceeds stereospecifically with respect to the reaction termini. It is therefore important to note that stereospecificity may not necessarily be indicative of a concerted reaction.

3.3   βγ-Unsaturated carbonyl compounds[13–17]
βγ-Unsaturated carbonyl compounds, though formally non-conjugated, can exhibit remarkable ultraviolet properties. Bicyclo[2,2,1]-hept-5-en-2-one (23) has an absorption maximum, in ethanol, at 308 nm ($\varepsilon =$

(23)   (24)   Ar = aryl

28 m^2 mol^{-1}) while the 5-aryl derivatives (24) have extinction coefficients up to 760 m^2 mol^{-1} for the n–π* band at *c.* 315 nm. The shift in the carbonyl absorption to longer wavelengths, and its enhancement in absorption over that of saturated ketones, is thought to result from the mixing of the n–π* and π–π* transitions. The photochemistry of these systems might therefore be expected to reflect something of this interaction.

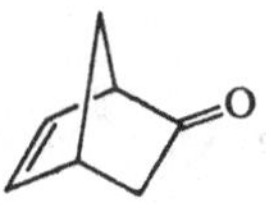

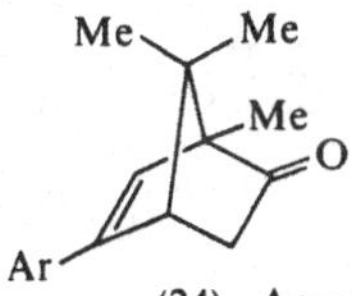

As well as undergoing reactions characteristic of the separate chromophores (geometrical isomerisation (section 2.1), dimerisation of the alkene (section 4.1), oxetane formation (section 4.2), reduction (section 5.3), Norrish Type I and Type II reactions of the carbonyl group (section 3.1))

$\beta\gamma$-unsaturated ketones exhibit two reactions that depend upon the bichromophoric interactions. These reactions can be described as [1,2] and [1,3] acyl shifts (cf. section 2.4) and are illustrated schemically below.[13-15] The [1,2] sigmatropic reaction is also known as the oxa-di-π-methane

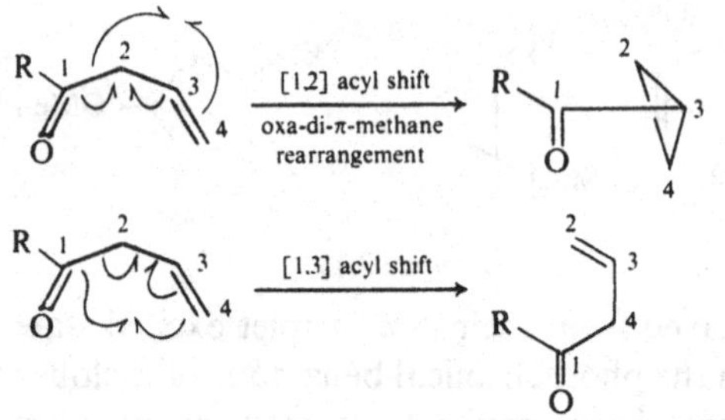

rearrangement and is analogous to the di-π-methane rearrangement of 1,4-dienes discussed in section 2.3. The reaction involves the migration of the acyl group from C_2 to C_3 (termed a [1,2] acyl shift), and formation of a bond between C_2 and C_4. The [1,3] acyl shift involves migration of the acyl group from C_2 to C_4 and relocation of the double bond between C_2 and C_3 (see section 2.4).

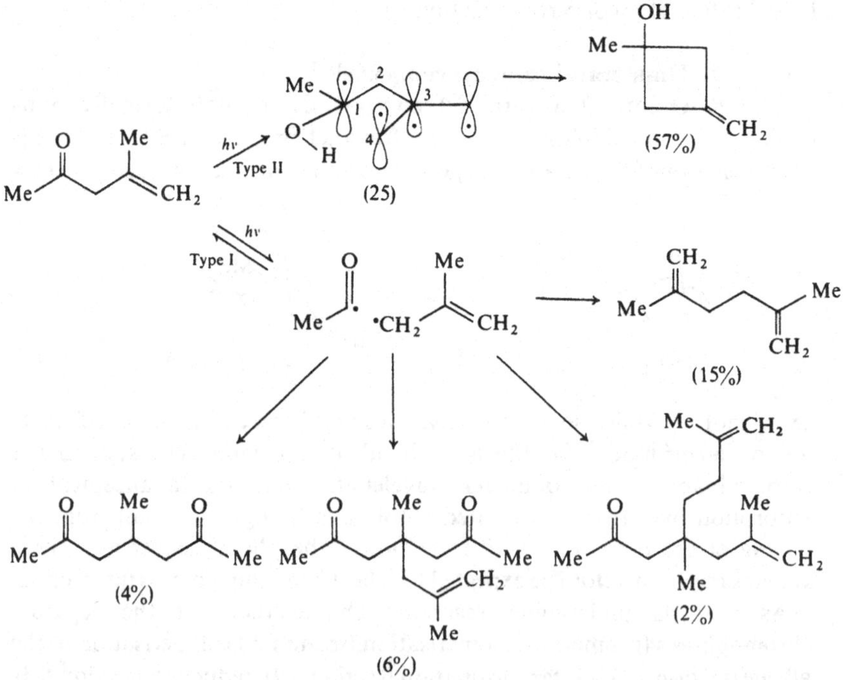

Scheme 3.16

Solution-phase photolysis of 4-methylpent-4-en-2-one (scheme 3.16) illustrates some of the more common Norrish Type I and Type II photochemical processes observed for carbonyl compounds. For this simple substrate the [1,2] shift product is not observed and this is a common feature with substrates which can geometrically isomerise thereby dissipating triplet energy which might otherwise be used to effect the oxa-di-π-methane rearrangement. A [1,3] acyl shift cannot be detected for this substrate since the product of the process is identical to the starting enone. The minor products of the reaction (scheme 3.16) are derived from Type I α-

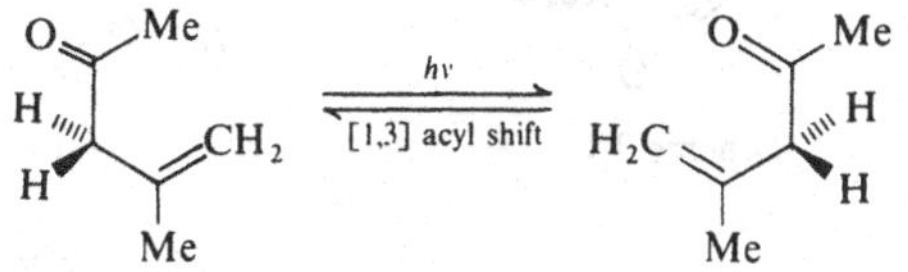

O—Me $\quad$ H,,, H $\quad$ CH$_2$ $\quad$ Me $\qquad$ $\xrightarrow[\text{[1,3] acyl shift}]{hv}$ $\qquad$ O—Me $\quad$ H$_2$C= $\quad$,,,H H $\quad$ Me

cleavage and subsequent reorganisation and recapture of the radical fragments. Cyclobutanol, the Norrish Type II product, is the major product of the reaction which proceeds from the n–π* singlet or from an exceedingly short-lived triplet state; the quantum yield is comparable with those measured for saturated ketones. The other products expected from fragmentation of the 1,4-diradical, namely acetone and allene, were not detected. The delocalised allylic end of diradical (25) (scheme 3.16) is constrained from adopting a conformation where the terminal orbital can overlap with the C_2–C_3 bond as required for fragmentation to occur. The introduction of two methyl substituents adjacent to the carbonyl group

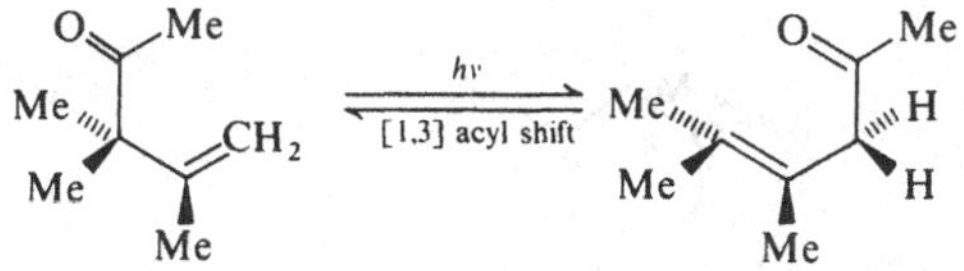

O—Me $\quad$ Me,,, Me $\quad$ CH$_2$ $\quad$ Me $\qquad$ $\xrightarrow[\text{[1,3] acyl shift}]{hv}$ $\qquad$ O—Me $\quad$ Me,,, ,,,H Me $\quad$ H $\quad$ Me

destroys the symmetry of the enone and a product resulting from a [1,3] acyl shift is observed. The product may result from a concerted shift of the acyl group (see section 2.4) or from α-cleavage and radical recombination in the cage containing the acyl and allylic radicals.

There are many examples of photochemical isomerisations of βγ-enones which proceed with [1,3] acyl migration as shown overleaf. The βγ-enone (26) formed by photochemical disrotatory ring closure of eucarvone (scheme 3.17), undergoes acyl migration to give (27). Similarly, photolysis of 2,2-dimethylcyclohept-3-enone (28) leads, by a [1,3] sigmatropic shift, to a five-membered ring ketone. This isomerisation is preferred to aldehyde formation, which would involve a Norrish Type I α-cleavage and hydrogen transfer through a five-membered cyclic transition state structure. In the corresponding cyclooctenone, hydrogen transfer can

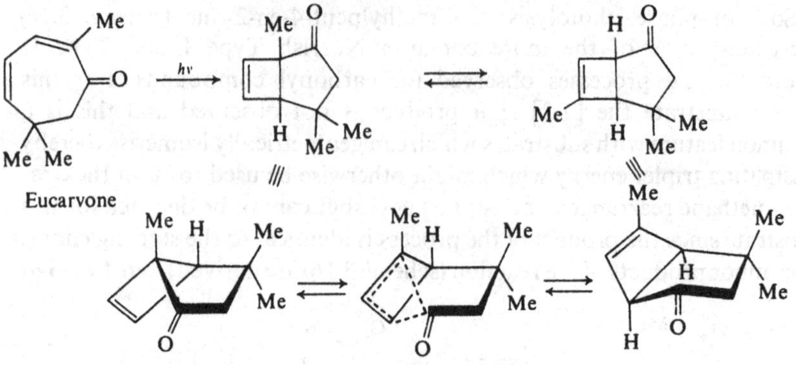

(26)   Scheme 3.17   (27)

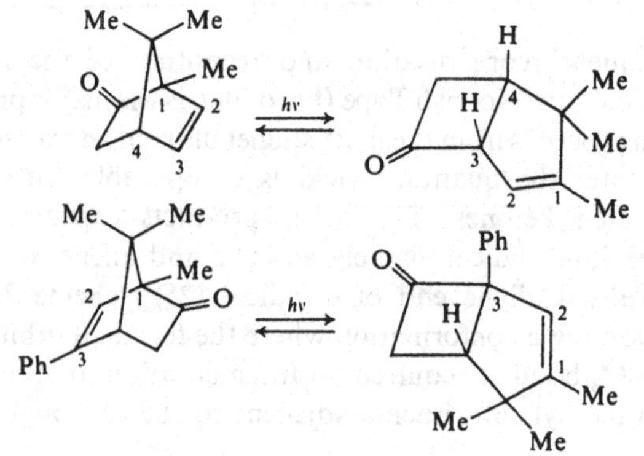

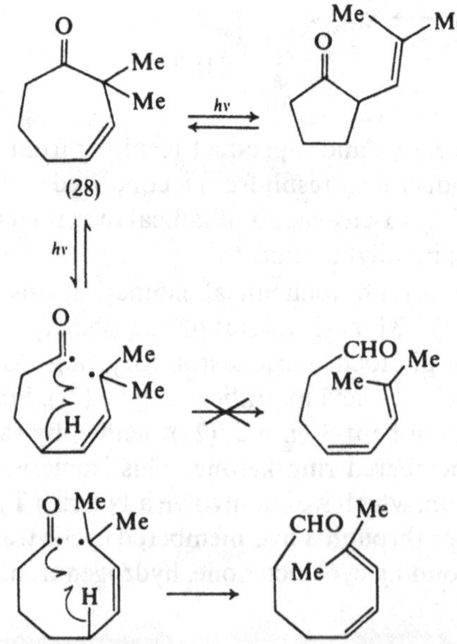

(28)

occur through a six-membered transition state structure and the aldehyde is formed to an extent of 10%.

The oxa-di-π-methane rearrangement[12] of βγ-unsaturated carbonyl compounds is most commonly observed in rigid systems where isomerisation of the alkene is inhibited. It occurs from the lowest π–π* triplet state and can be sensitised by typical triplet sensitisers. When the reactive triplet is populated by direct irradiation reaction is inhibited by typical triplet quenchers such as dienes. *Bona fide* phosphorescent emission from an enone chromophore has not been observed and so the energy of the excited state is not well established. Studies using sensitisers of varying triplet excitation energy suggest that the energy of the triplet state is about 74–78 kJ mol^{-1} above that of the ground state. The n–π* excited singlet is estimated to be 86–87 kJ mol^{-1} above ground state (S_0) and the n–π* triplet within 2 kJ mol^{-1} of the acetone triplet which is at 78 kJ mol^{-1}.

The bicyclic ketone (29) is a relatively rigid βγ-enone and on direct or sensitised photolysis the parent compound (29; R = H) gives only tricyclic

cyclopropyl ketone. The subtlety of the effects of small changes of structure on reaction course is demonstrated by methyl substitution on C_2. The methyl ketone (29; R = Me) gives both the [1,2] and [1,3] acyl shift products on direct irradiation, but only the [1,2] product from a triplet sensitised photolysis. The corresponding dimethyl ketone (30), however, gives only the [1,3] product on direct photolysis and only the [1,2] product on triplet sensitisation. The substituents may affect the population and energy of the various excited states, and at the same time affect the efficiency

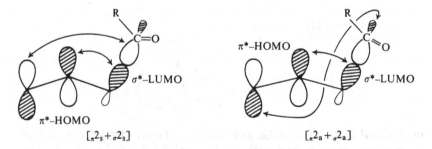

(30)

of interconversion mechanisms between these states. In addition, the substituents will affect the partitioning of an excited state between the various products that are potentially possible from that state. Thus several factors remain to be unravelled before an understanding of the excited state chemistry of these systems is possible.

The [1,3] acyl shift is generally observed on direct photolysis of a $\beta\gamma$-enone and, as such, is not affected by typical triplet quenchers. This indicates that the reaction occurs from the singlet excited state. Nonetheless, there are reports of [1,3] acyl shifts which occur from a triplet excited state. However, for cases where this has been observed it is not known whether the triplet is the same triplet excited state that is responsible for the [1,2] acyl shift. A [1,3] shift from the singlet state is a reaction which competes with ISC to the T_1 excited state and hence with the [1,2] acyl shift. The competitiveness of these processes is demonstrated by the partitioning shown in the foregoing examples to give [1,2] products, a mixture of [1,2] and [1,3] products, and [1,3] products.

A [1,3] acyl sigmatropic shift can be described as a $[_{\pi}2_s + _{\sigma}2_s]$ reaction and as such is orbital symmetry allowed[16] as has been discussed for sigmatropic reactions (section 2.4). It is not reasonable to consider the

$$[_{\pi}2_s + _{\sigma}2_s] \qquad\qquad [_{\pi}2_a + _{\sigma}2_a]$$

alternative allowed photochemical description $[_{\pi}2_a + _{\sigma}2_a]$ since this would require the acyl group to migrate across the nodal plane of the carbon–carbon double bond. Even allowing for the possibility of twisting of the double bond in the excited state the geometry is too constrained for the migration to occur.

In a similar way the [1,2] acyl shift can be considered as a $[_{\pi}2_s + _{\sigma}2_a + _{\pi}2_s]$

or a $[_\pi 2_a + _\sigma 2_a]$ process in direct analogy with the di-π-methane analogue (section 2.4) and as such can be considered as orbital symmetry allowed from an excited state. However, the reaction is known to occur from an

$$[_\pi 2_s + _\sigma 2_a + _\pi 2_s] \qquad [_\pi 2_a + _\sigma 2_a]$$

excited triplet state and the appropriateness of this treatment for a reaction which requires an intermediate with a lifetime of sufficient length for spin inversion to the ground state to occur is of some concern.

Photolysis of the $\beta\gamma$-enone (31) is of interest[17] since [1,2] and [1,3] acyl shift products are observed from both direct and sensitised photolysis. The

Product ratio			
hv direct	2.73	:	1
hv sens	0.031	:	1
Δ	0.7	:	1

[1,3] shift product is the major product on direct photolysis and the C$_5$-epimeric [1,2] shift products are the major products from sensitised photolysis. The n–π^* excited triplet state of the starting ketone can be independently produced from thermal fragmentation of the dioxetane (32) (see section 5.1). Fluorescence studies show that only a small amount (15%) of the S_1 state is produced from the thermolysis and hence it can reasonably be assumed that the dominant excited state is the n–π^* triplet. Thus the n–π^* triplet gives a more even distribution of the [1,2] and [1,3] products than is obtained from either the direct or sensitised photolysis reactions. Quenching experiments show unequal quenching of the [1,2] and [1,3] processes in the sensitised reaction; this is consistent with the two products

arising from different triplet states, the [1,2] shift occurring from the lower energy $\pi-\pi^*$ T_1 state and the [1,3] shift from T_2, the $n-\pi^*$ triplet. Molecular orbital calculations suggest that the $n-\pi^*$ singlet excited state has a tendency to undergo α-cleavage and subsequent formation of a bond between the carbonyl and C_1. By comparison the $\pi-\pi^*$ triplet excited state does not exhibit significant weakening of the α-bond, but shows interaction between the carbonyl carbon and C_2. This is indeed the molecular change required for a [1,2] shift. The [1,3] acyl shift is at least in part, a radical process proceeding via a cage radical pair. The cage radical pair is formed predominantly from the singlet excited state S_1 ($n-\pi^*$) at temperatures of $c.$ 50°C and from a triplet state T_2 ($n-\pi^*$) at lower temperatures ($c. -50$°C). The radical pair is also capable of disproportionating to acetaldehyde and diene. The dominant reaction is however from S_1 which gives [1,3] acyl shift product in competition with ISC and fluorescence. ISC gives both T_2, which undergoes [1,3] acyl shift, and T_1 which undergoes [1,2] acyl shift.

The question as to whether the acyl group in a [1,3] shift reaction remains bound at all times as the reaction proceeds or whether α-cleavage occurs and the radical fragments recombine is a matter of some interest. In a few unsensitised reactions of $\beta\gamma$-enones, the acyl radical fragment has been isolated as the dimer, for example butanedione from (33). The presence of

(33)

allyl radical has similarly been confirmed by isolation of the dimer (35), in the photolysis of (34a) (scheme 3.18). So for these two reactions a proportion of the products must arise by way of α-cleavage and loss of the radicals from the radical cage to allow their subsequent dimerisation. For the parent enone (34a) the [1,3] reaction is degenerate and can only be detected if an optically active starting enone is involved. As the reactions of (34a,b) proceed the optical activity of the enone decreases as a racemic mixture is formed. Labelling experiments with the deuterated compounds (34b) and (34c) have also shown that the [1,3] shift is intramolecular since photolysis of a mixture of the two compounds does not result in any crossover product with $D_{(6)}$ labelling. However, this does not prove that the reaction is concerted; it is possible for α-cleavage to occur with subsequent reorganisation and recombination within a tight radical cage. On direct photolysis there is no [1,2] product detected and ISC from the singlet excited state to the triplet does not compete with the [1,3] shift (scheme 3.18).

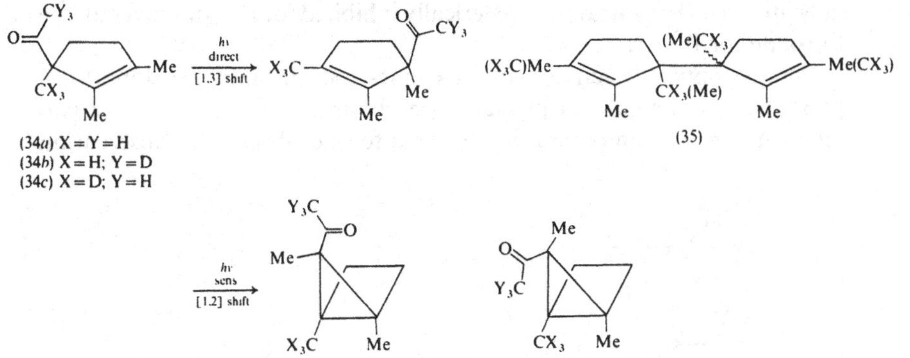

Scheme 3.18

The intramolecular nature of the [1,3] shift is further supported by the stereospecificity observed in the direct photolysis of the βγ-enone (36) and its steroidal analogue (37) (scheme 3.19). An aldehyde is also formed from the enone (36) by non-stereospecific H(D)-transfer to the acyl radical when $R^3 = H$ and from a stereospecific H(D)-transfer when $R^3 = $ methyl. This

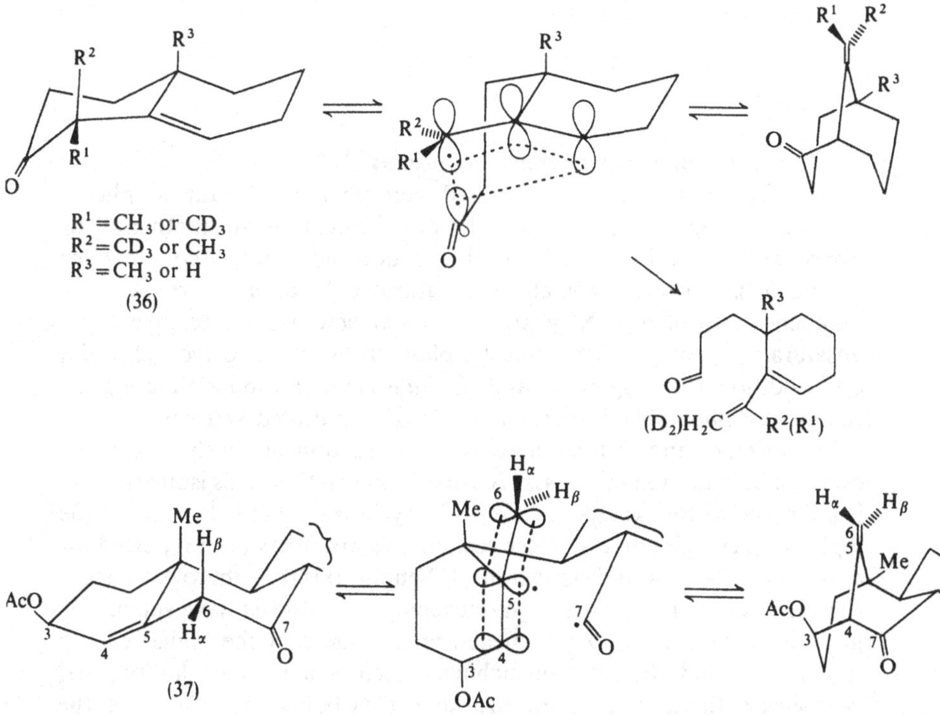

$R^1 = CH_3$ or CD_3
$R^2 = CD_3$ or CH_3
$R^3 = CH_3$ or H

(36)

Scheme 3.19

indicates that the radical pair is sterically inhibited for R^3 = methyl but free to rotate for R^3 = H.

For the optically active benzoyl enones racemisation, resulting from [1,3] shifts, and crossover products are obtained when a mixture of two different benzoyl compounds is subjected to photolysis. This observation

along with *chemically induced dynamic nuclear polarisation* (CIDNP) n.m.r. studies shows the presence in the reaction of radical species. CIDNP n.m.r. studies of the acyl analogues did not, however, indicate the presence of radical intermediates.

3.4   αβ-Unsaturated carbonyl compounds[13,16]

Many simple αβ-unsaturated ketones are resistant to photo-chemical change in inert solvents but irradiation in methanol or isopropanol results in photochemical reduction and reductive coupling (see section 5.3). However, not all αβ-unsaturated ketones are resistant to photochemical change. Migration of the alkenic π-bond to give a βγ-unsaturated ketone, dimerisation by photochemically-allowed $[_\pi 2_s + _\pi 2_s]$ cycloaddition, rearrangement, and *cis–trans* isomerisation are among the commonest photochemical reactions of αβ-unsaturated ketones.

The primary and fastest photochemical reaction of acyclic and some cyclic αβ-unsaturated ketones is *cis–trans* isomerisation. This isomerisation is not expected for cyclopent-2-enone or cyclohex-2-enone because of the high strain energies involved (see section 2.1), but it has been reported for *cis*-cyclooct-2-enone and *cis*-cyclohept-2-enone. It is thought to occur from the triplet excited state although quenching experiments have sometimes given negative results. Diene quenchers can intercept the higher energy triplet state and therefore quenching, which is a bimolecular process, competes with twisting about the carbon–carbon double bond of the excited triplet (see section 2.1). Flash spectroscopic studies have led to the

identification of two transient species from cyclohept-2-enone. The longer
lived of the two (45 s in cyclohexane) is *trans*-cyclohept-2-enone. In
methanol the lifetime of this compound is dramatically reduced (0.03 s)

because of *syn*-addition of the solvent to the strained double bond. The
second transient of much shorter lifetime in cyclohexane has not yet been
identified. The photochemically induced addition of methanol across the
carbon–carbon bond of cyclohept-2-enone and cyclooct-2-enone has been
established as *trans*, but is more complex since it involves initial
isomerisation to the *trans*-cycloalkenones followed by *syn*-addition of
methanol across the strained double bond.

Isomerisation to a βγ-unsaturated ketone, structural rearrangement and
intermolecular dimerisation and addition may compete with or follow after
geometrical isomerisation. In suitably constrained cases the excited alkene
can abstract a distal hydrogen atom as shown in scheme 3.20; the reaction
probably involves the triplet π–π* excited state.

Photolysis of the hexanones (38) and (39a) results in migration of the
carbon–carbon double bond to give the βγ-unsaturated isomers. For (38)
photolysis results in an (E)/(Z) mixture of isomeric products while for the 5-
methyl analogue (39a) only one βγ-isomer is possible. The reaction involves
the intramolecular abstraction of a hydrogen atom from the γ-position

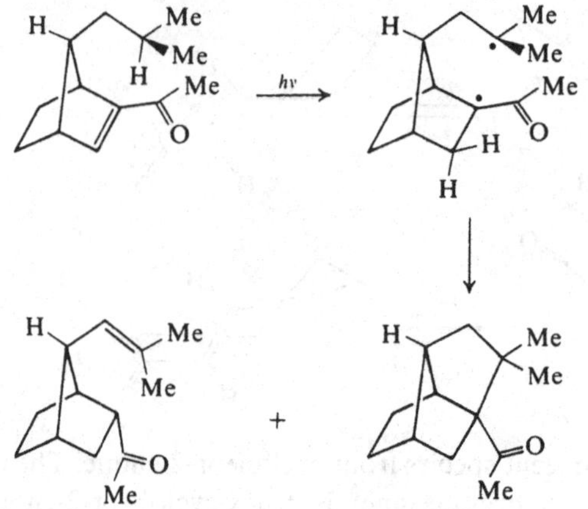

Scheme 3.20

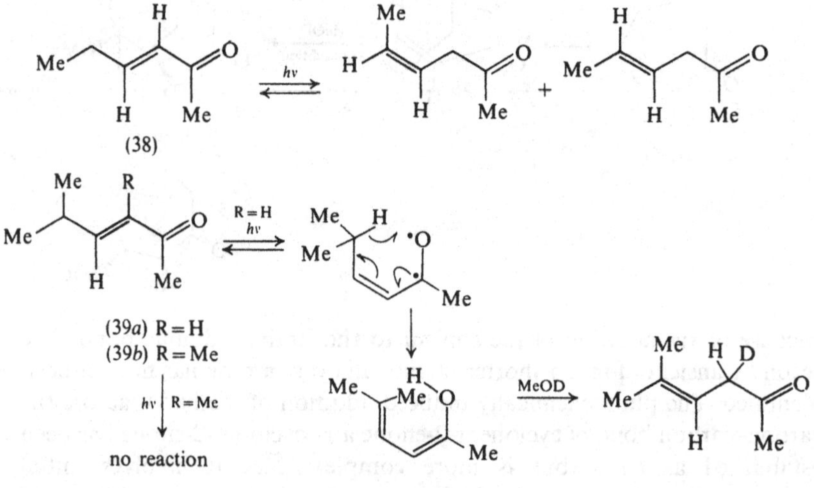

Scheme 3.21

through a cyclic six-membered transition state structure as shown in scheme 3.21, and proceeds from the n–π* excited triplet state. If the reaction is carried out in d¹-methanol as solvent, the resulting enol–ketone isomerisation results in deuterium incorporation at C_3. The reaction sequence is identical with Norrish Type II bond cleavage (section 3.1) except that π- rather than σ-bond cleavage occurs. It is perhaps surprising that 3,5-dimethylhex-3-en-2-one (39b), is unreactive to light. However, the presence of the additional methyl group lowers the energy of the π–π* triplet below that of the n–π* triplet state. Aromatic carbonyl compounds also have low-lying triplet states and are similarly resistant to photochemical hydrogen transfer reactions. The failure of other simple αβ-unsaturated pentenones, such as mesityl oxide (4-methylpent-3-en-2-one), to undergo photochemical change despite the availability of a γ-hydrogen atom, may be a consequence of the additional energy required to abstract a primary hydrogen atom.

The conjugated 1,3-dienol intermediate in the isomerisation of αβ- to βγ-unsaturated ketones has been isolated from 1-acetylcyclooctene (40). The

αβ-unsaturated ketone is thought to undergo initial *cis–trans* isomerisation; the carbonyl oxygen atom is then suitably located to abstract the γ-hydrogen atom. The unusual stability of these dienols results from the relatively rigid s-*trans* diene structure and from the geometry of the eight-membered ring. In general dienol intermediates undergo rapid ketonisation with loss of conjugation.

The αβ-unsaturated ketone (41) gives the βγ-enone but this reaction also involves skeletal rearrangement. The stereochemistry at the migrating

(41) R = D or Ph

centre is lost when R = D, indicating that for these substrates a diradical intermediate is most likely involved in the reaction.

When an αβ-unsaturated ketone lacks a γ-hydrogen atom, photolysis usually results in an intramolecular δ-hydrogen transfer through a seven-membered cyclic transition structure.

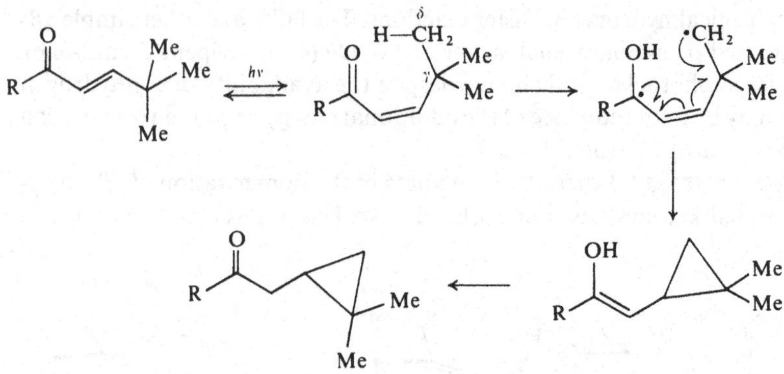

Cycloadditions of cyclic $\alpha\beta$-unsaturated ketones are generally intermolecular and of the $[_\pi 2 + _\pi 2]$ type. Often these involve photoinduced geometrical isomerisation followed by a ground state reaction of the highly reactive *trans*-cyclic enone (see section 4.4). Intramolecular cycloadditions are also known and an example is given in scheme 3.22.

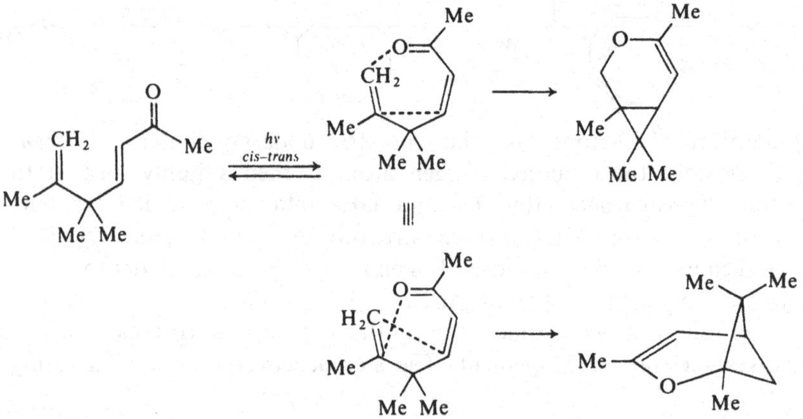

Scheme 3.22

Skeletal rearrangement can occur from excited enones. For example, testosterone acetate (42) on photolysis gives a wide range of products the

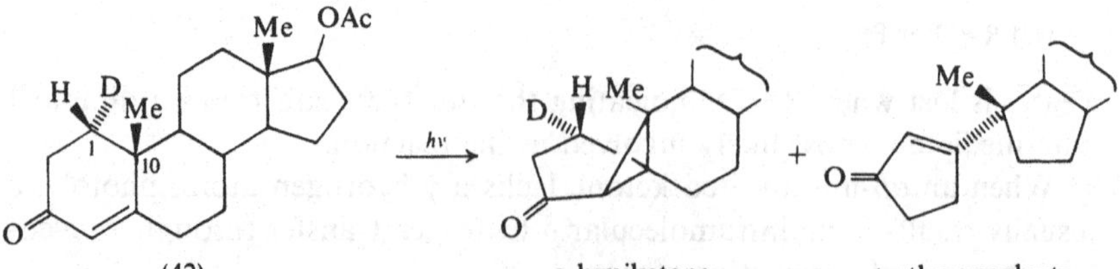

(42)   a lumiketone   +other products

nature of which is very solvent dependent. In butanol two rearranged compounds are of particular interest. The first of these is known as a lumiketone and the rearrangement is often referred to as the lumiketone rearrangement. Deuterium labelling experiments have shown that the reaction proceeds with retention of configuration at C_1 and inversion at C_{10}. The reaction has all the characteristics of a concerted $[_\pi 2_a + _\sigma 2_a]$ reaction,[16] allowed in the excited state and illustrated by (43). An

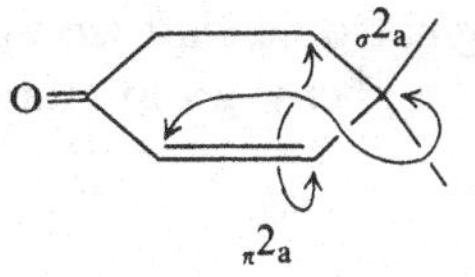

(43)

analogous lumiketone rearrangement occurs for the optically active enone (44) which affords the two optically pure diastereoisomeric lumiketones resulting from inversion of configuration at C_4. For this substrate, reaction

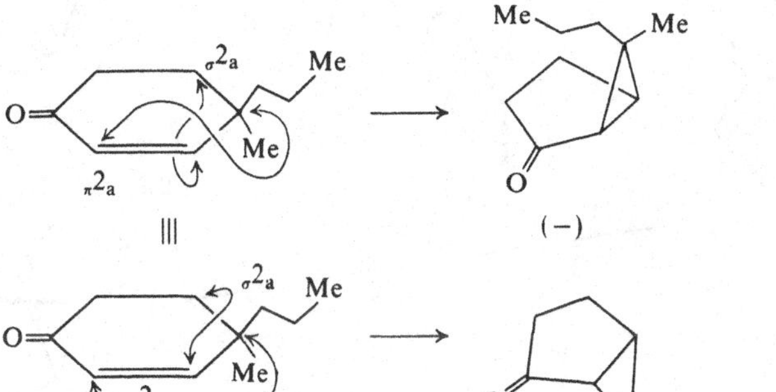

(44)

can occur at both faces of the enone whereas the constraints of the fused B and C rings in testosterone acetate (42) prevent this happening. In addition to cyclopropyl ketones, enones (42) and (44) also give a cyclopentenone product but its mode of formation is not well understood. The optical activity of the cyclopentenone suggests that the reaction could be represented as a $[_\sigma 2_a + _\sigma 2_a]$ reaction with migration of the hydrogen at C_3 to C_4 leading to inversion of configuration at C_4, a process occurring synchronously with ring contraction. However, this description is simplistic since it does not fully describe the energy changes that must occur from initial excitation of the enone. A further constraint on the lumiketone rearrangement is a requirement for 4,4-disubstitution which consequently limits the utility of the reaction. Furthermore, with a vinyl or aryl substituent at C_4 a reaction which has analogy with the di-π-methane rearrangement (section 2.3) and the [1,2] sigmatropic shift (section 2.4) occurs. This reaction involves the migration of a phenyl group from C_4 to

C_3 and closure between C_2 and C_4. The formation of the major product is rationalised in terms of the bridged intermediate showing inversion at C_4.

3.5   Cyclohexadienones[16,18,19]

It is now more than 150 years since the first photochemical reaction of a cyclohexadienone chromophore in the sesquiterpene α-santonin was noted. However, it is only in recent times that the structures of the products have been determined. α-Santonin undergoes photochemical rearrangement in non-nucleophilic media to give the cyclopropyl ketone lumisantonin, which is itself photochemically labile and yields the linearly conjugated dienone, mazdasantonin. In nucleophilic media, water or ethanol, mazdasantonin gives the ring-cleaved acid or ester (45) and this compound is the ultimate product when α-santonin is photolysed in water

α-santonin   lumisantonin   mazdasantonin

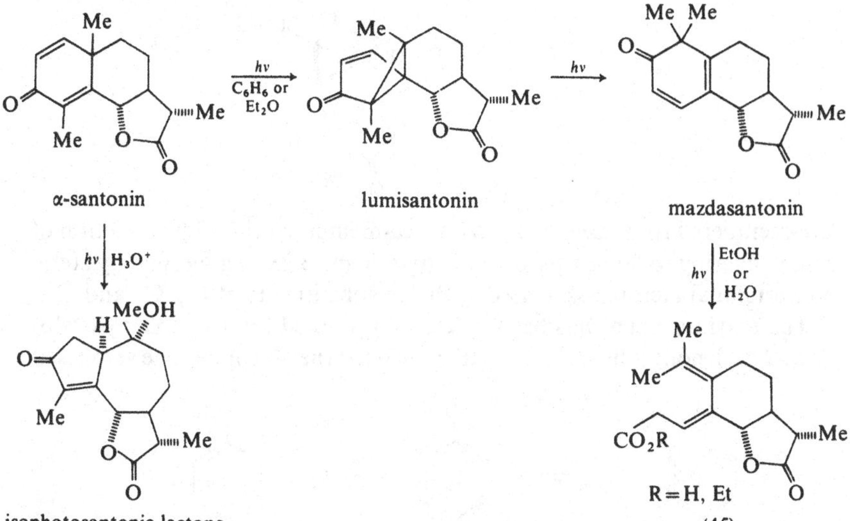

isophotosantonic lactone

R = H, Et

(45)

or ethanol. A different product, isophotosantonic lactone, is obtained when the reaction is carried out in aqueous acidic media. This hydroxyketone arises from a cyclopropyl intermediate and the details of the reaction are discussed below.

The primary photochemical reaction of a cyclohexadienone, and that observed for α-santonin by its conversion to lumisantonin, can be formally represented as a $[_{\pi}2_a + _{\sigma}2_a]$ reaction[16] shown schematically in scheme 3.23. This representation of the rearrangement does not take into account

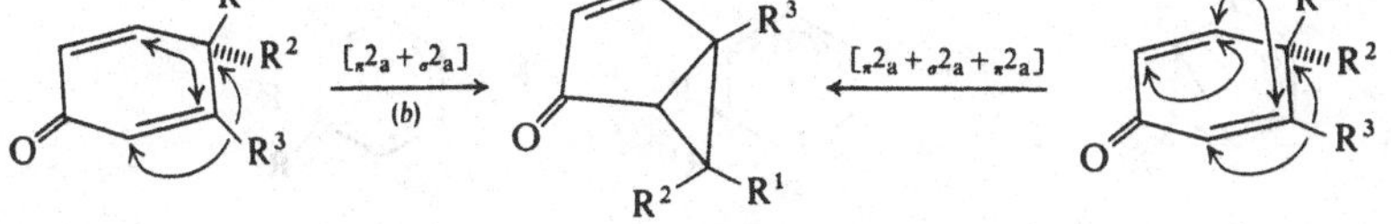

Scheme 3.23

the second double bond of the dienone and perhaps the reaction should more correctly be considered as a $[_{\pi}2_a + _{\sigma}2_a + _{\pi}2_a]$ process (scheme 3.23). The alternative symmetry-allowed reaction course, that which involves suprafacial addition to the double bond of the cyclohexadienone, a $[_{\pi}2_s + _{\sigma}2_s]$ process, would result in an impossible stereochemistry with a five-membered ring *trans*-fused to a three-membered ring. The direction of ring puckering to favour path (*a*) *vs* (*b*) (scheme 3.23) is governed by steric and torsional factors associated with the substituents at C_3, C_4 and C_5.

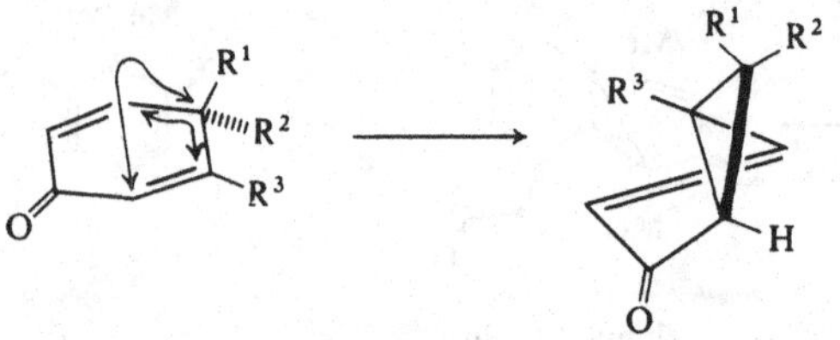

The rearrangement has been extensively studied for 4,4-diphenylcyclo-hexa-2,5-dienone where the quantum yield for the acetophenone sensitised

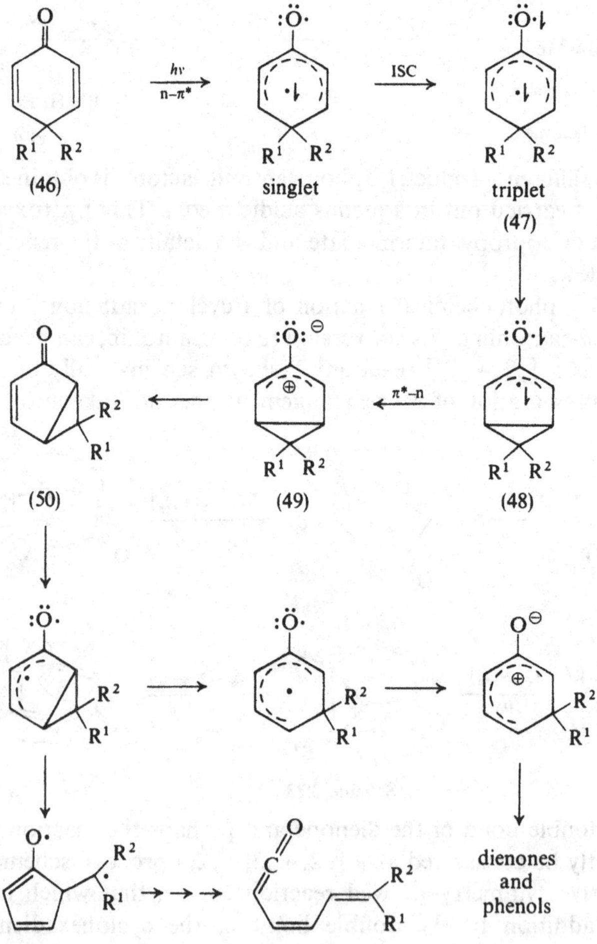

Scheme 3.24

and unsensitised reactions is high and the same ($\Phi = 0.85$). This implies that the same excited triplet state is involved in both reactions since it reflects an efficient ISC (S_1-T_1) in the unsensitised reaction. In all reported cases the reaction is believed to involve the n–π* triplet. However the varied and relatively low efficiency with which the energy of the triplet excited state of cyclohexadienones can be transferred to the quenchers naphthalene, 1,3-pentadiene, and 1,3-hexadiene casts doubt on the precise nature of the quenchable triplet; it could be an n–π* or a π–π* triplet. Zimmerman has attempted to rationalise the reaction course of the excited state dienone by use of simple Lewis structures and molecular orbital representations. For the rearrangement of 4,4-disubstituted cyclohexadienone (scheme 3.24), the suggested mechanism involves C_3–C_5 bonding from the excited triplet state (47). An examination of the molecular orbitals of the n–π* triplet state of the dienone indicates that an increase in C_3–C_5 bonding results from an electron occupying the lowest π*-molecular orbital (fig. 3.1, p. 114). In this mechanism, the excitation energy gained in going from the ground state (46), to the excited triplet state (47), is lost after C_3–C_5 bonding (48), to give the zwitterion (49), which subsequently rearranges to the product 6,6-disubstituted bicyclo[3,10]hex-3-en-2-one (50). The product (50) is itself photochemically labile and undergoes further rearrangement as shown in scheme 3.24.

The evidence for zwitterion (49) being an intermediate in these reactions follows from the rearrangement of the independently generated ground state zwitterion (52) to the same products as are formed in the photochemical reactions. It is significant that the zwitterion (52) rearranges

Fig. 3.1. Molecular orbitals of cyclohexadienone showing n–π^* excitation.

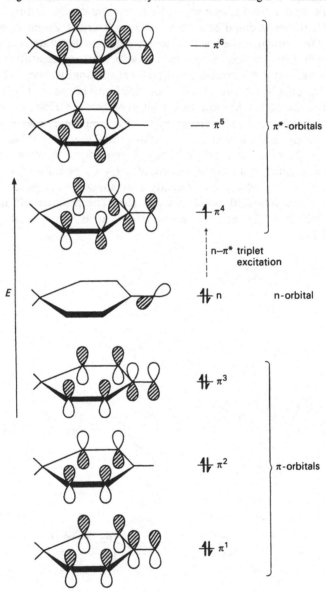

efficiently to the cyclopropyl ketone (53) rather than reverting to the starting dienone (51). Moreover, the stereospecificity observed in these rearrangements must be accounted for if a zwitterion is involved. In fact, when the zwitterion (54) is generated thermally in its gorund state from unsymmetrical diarylbicyclo[3,1,0]hexanone with potassium *t*-

(54)

butoxide, it rearranges stereospecifically as shown. A pivot mechanism involving C_1–C_6 bond cleavage, rotation about the C_5–C_6 bond, and C_4–C_6 bond formation is excluded because this would result in formation of the wrong product. The preferred mechanism must involve either successive migrations of C_6 from C_5 to C_4, and from C_1 to C_5, with retention of configuration at C_6 at both stages, or a [1,4] shift of C_6 from C_1 to C_4 with concomitant inversion of configuration at C_6. The zwitterion thought to be

a precursor to (56) in the photolysis of dienone (55) has been trapped in acidic alcoholic media thereby adding further support to the involvement of an ionic intermediate in the photolysis of such dienones.

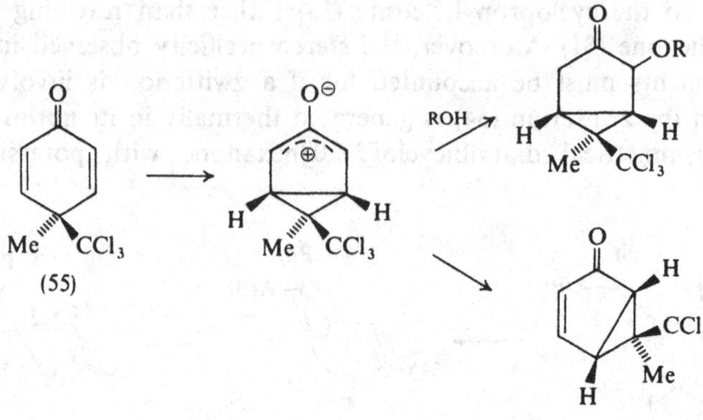
ROH
Me
CCl$_3$
H
OR

(55)

(56)

A most significant feature of these dienone rearrangements is the degree of stereoselectivity in the reaction. The product stereochemistry has been determined by X-ray crystallographic methods for a series of α-santonin-like compounds. The assignment of photoproducts in most other systems (scheme 3.25) has generally been made by analogy with the photochemical products from α-santonin. The reaction course is influenced by steric constraints. Thus for the 1-dehydro-B-nortestosterone acetate (57) the

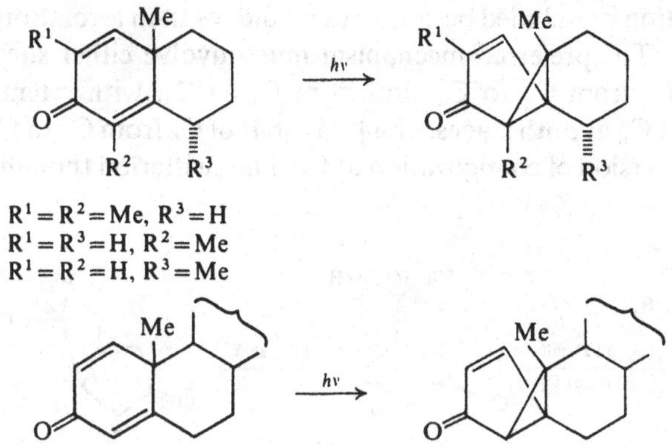
$h\nu$
Me
$R^1=R^2=$Me, $R^3=$H
$R^1=R^3=$H, $R^2=$Me
$R^1=R^2=$H, $R^3=$Me

Scheme 3.25

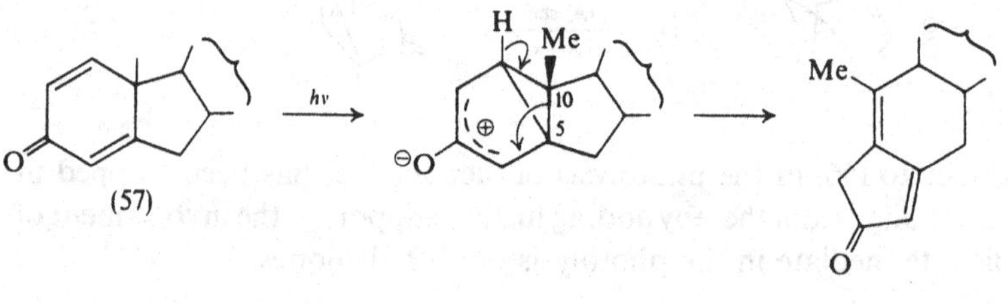
$h\nu$
H
Me
10
5

(57)

zwitterion intermediate collapses to a linearly conjugated ketone. In this instance cleavage of the C_5–C_{10} bond relieves the strain of the five-membered ring B of the norsteroid system.

Photolysis in aqueous acid media causes the reaction to take a different course. The initially formed cyclopropyl zwitterion is protonated and then cleavage of the three-membered ring gives hydroxyalkenes (scheme 3.26).

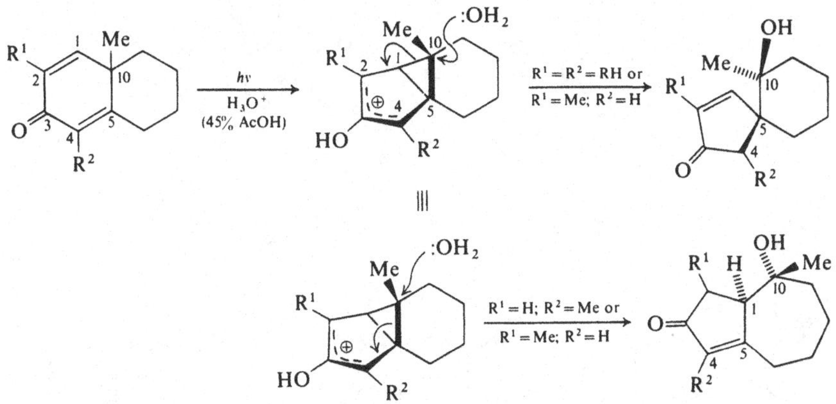

Scheme 3.26

The nature of the substituents R^1 and R^2 at C_2 and C_4 respectively, will influence the reaction course, in favour of either the spiran or the fused-ring ketone. Electron-donating substituents at C_4 will stabilise the reaction which proceeds by development of charge at C_4 and thus lead to the fused-ring ketone. An electron-donating substituent at C_2 will, in like manner, favour spiran formation. Substitution in the second ring can also markedly influence the course of the reaction.

The cyclopropyl ketones formed by photolysis of dienones in neutral solvents can themselves be regarded as dienones in which one double bond has been replaced by a strained cyclopropane ring. Not surprisingly, these ketones are photochemically labile and their reactions generally proceed through a dipolar species as shown in scheme 3.27. The dipolar intermediate (58) can rearrange by three routes. A 1,2-methyl migration affords linearly conjugated dienone (path (a), scheme 3.27) which, when $R^2 = H$, rapidly tautomerises to the phenol; a 1,2-alkyl migration results in the formation of a spiro-dienone (path (b)), and an alternative 1,2-methyl migration (path (c)) gives a new dienone. These dienone products can photochemically rearrange, thereby adding to the multiplicity and complexity of the reaction.

Lumisantonin, with a *trans*-fused lactone ring, is one of a limited number

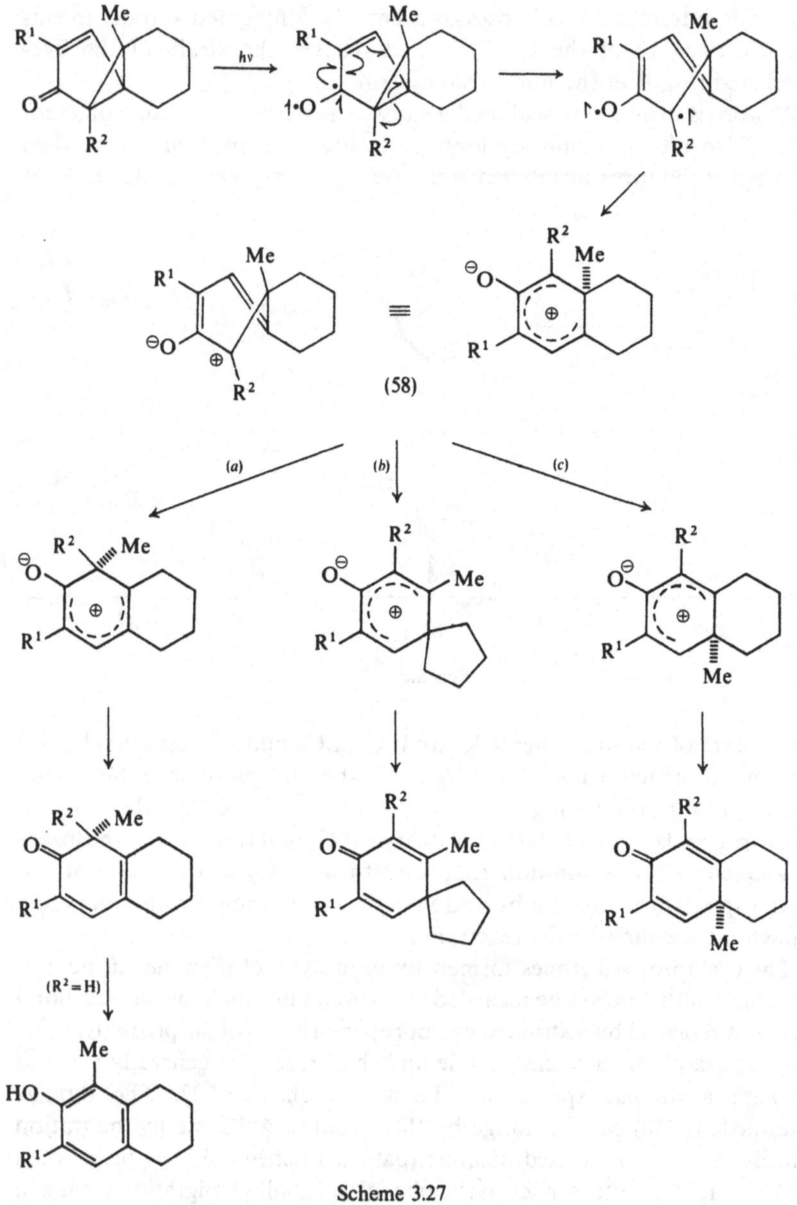

Scheme 3.27

of examples where the 1,2-methyl migration (path (a)) is the preferred reaction course. The isomer of lumisantonin with a *cis*-fused lactone ring (59) does not likewise rearrange but gives a spiran by the path (b) route. The absence of spiran products from lumisantonin is probably a reflection of the

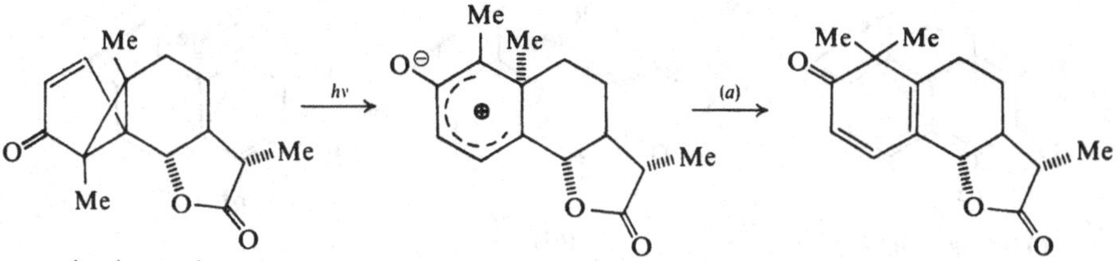

lumisantonin

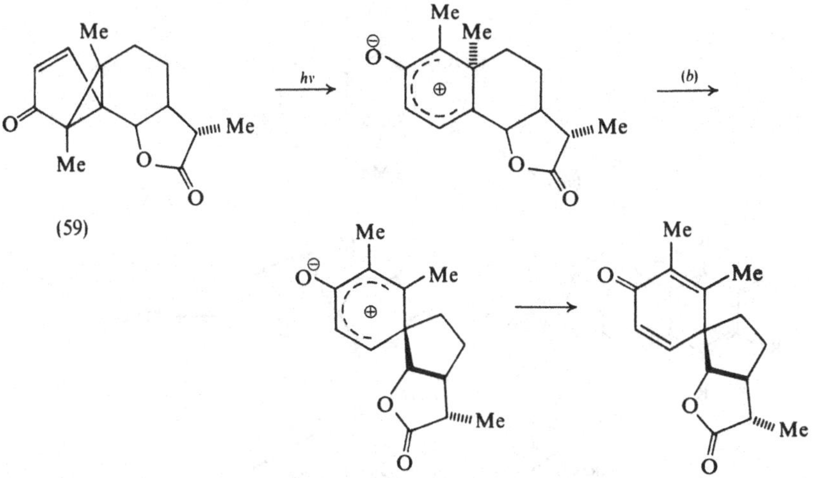

(59)

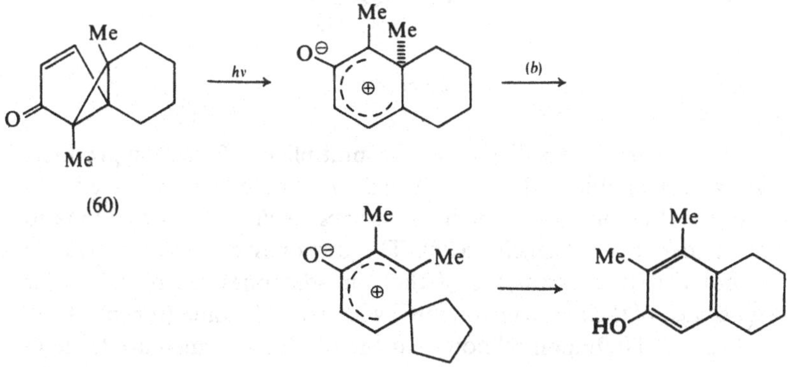

(60)

high energy of two *trans*-fused five-membered rings. The parent cyclopropyl ketone (60), in the absence of the fused lactone ring, rearranges (path (b)) to the spiro-dienone; further rearrangement leads to a phenol. The photochemistry of substituted testosterone acetate dienone derivatives

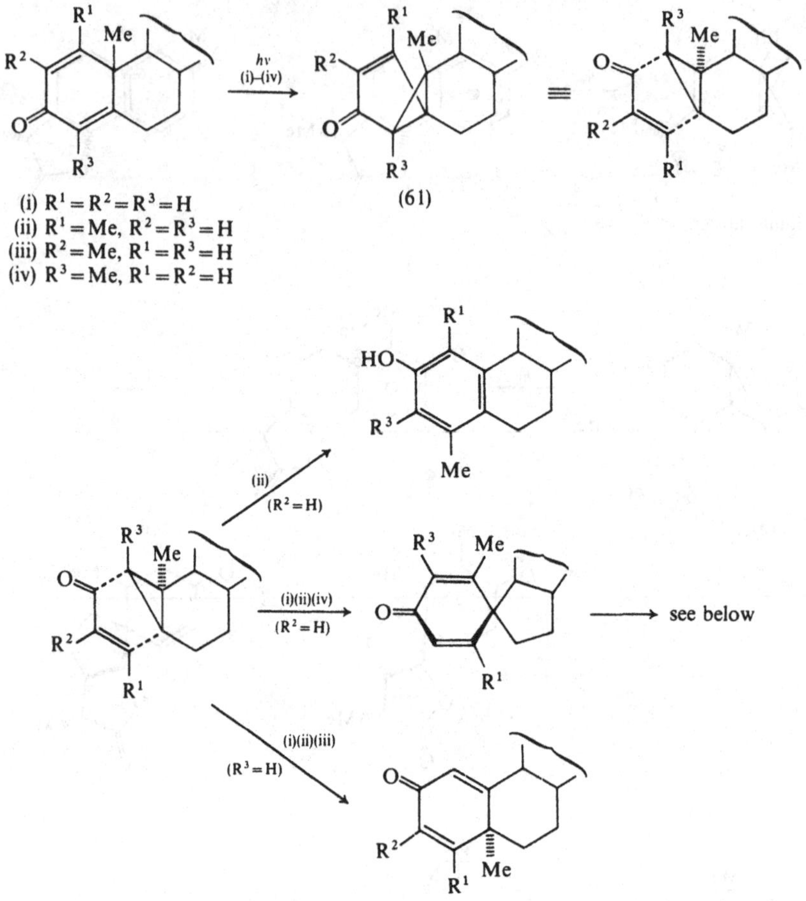
hν (i)–(iv)
(61)
(i) R¹ = R² = R³ = H
(ii) R¹ = Me, R² = R³ = H
(iii) R² = Me, R¹ = R³ = H
(iv) R³ = Me, R¹ = R² = H
(ii) (R² = H)
(i)(ii)(iv) (R² = H)
see below
(i)(ii)(iii) (R³ = H)

Scheme 3.28

is particularly instructive in displaying the multiplicity of reaction products and pathways available to dienones. The dienones are first converted into the cyclopropyl ketones (61) which themselves readily photorearrange to dienones and phenols (scheme 3.28). The dienones can each rearrange further, but the transformations of the spiro-dienones are of particular interest because of their conversion into cyclopropyl ketones (scheme 3.29).

Woodward & Hoffmann[16] note that the products formed are three of the four which are possible from a $[_\pi2_a + _\sigma2_a]$ cycloaddition reaction. The cyclopropyl ketones are themselves able to rearrange further to new dienones and phenols. The linearly conjugated diene so produced rearranges to phenols and, in aqueous solution, carboxylic acid derivatives also result from ketene intermediates (see scheme 3.24).

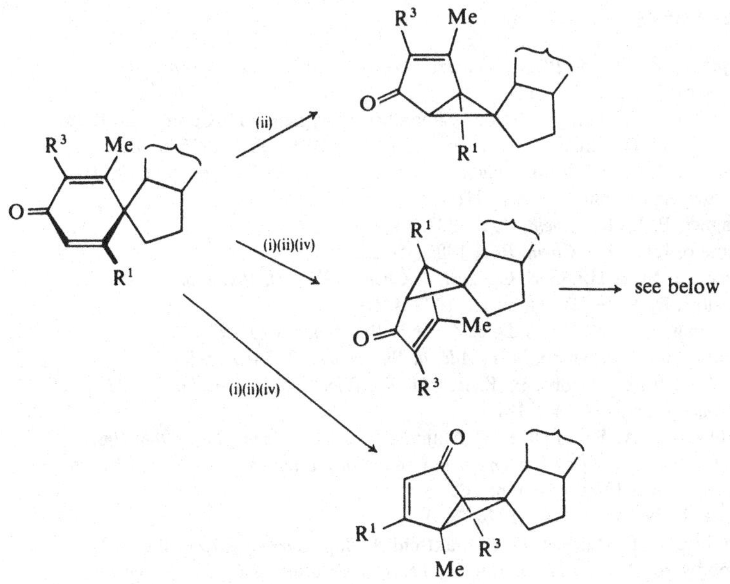

Scheme 3.29

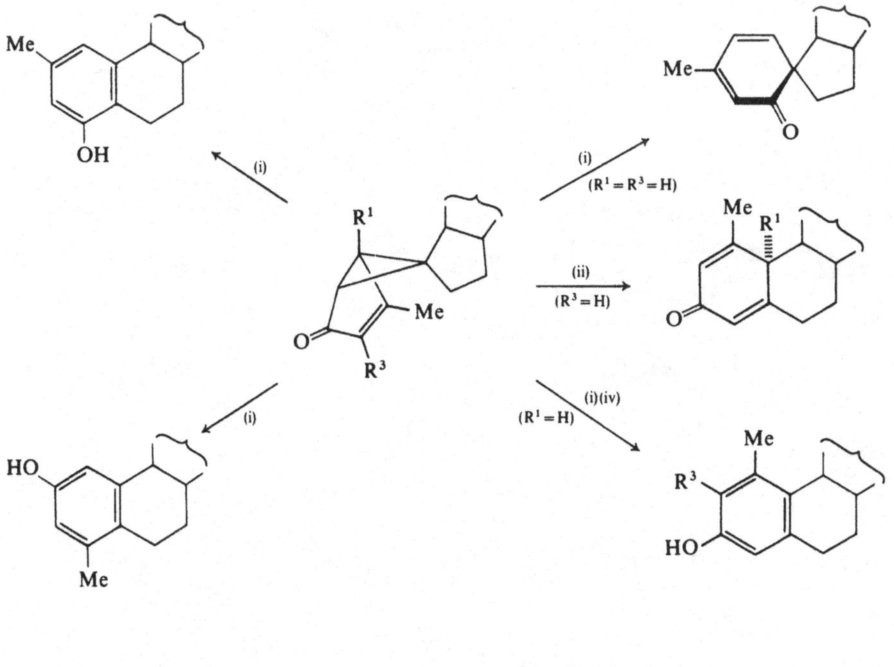

References

1. Bryce-Smith, D., ed., *Photochemistry – a Specialist Periodical Report*, The Chemical Society, Vols. 1–14.
2. Turro, N. J., Modern Molecular Photochemistry, Benjamin/Cummings, 1978.
3. Dauben, W. G., Salem, L. & Turro, N. J., *Acc. Chem. Res.*, 1975, **8**, 41.
4. Wagner, P. J., in *Rearrangements in Ground and Excited States*, ed. de Mayo, P., Academic, 1980, Vol. 3, p. 381.
5. Wagner, P. J., *Acc. Chem. Res.*, 1983, **16**, 461.
6. Scaiano, J. C., *Acc. Chem. Res.*, 1982, **15**, 252.
7. Coxon, J. M. & Hii, G. S. C., *Aust. J. Chem.*, 1977, **30**, 161, 835.
8. Breslow, R., *Chem. Soc. Reviews*, 1972, **1**, 553.
9. Chapman, O. L. & Weiss, D. J., *Organic Photochemistry*, 1973, **3**, 197.
10. Morton, D. R. & Turro, N. J., *Adv. in Photochem.*, 1974, **9**, 197.
11. Strohrer, W.-D., Jacobs, P., Kaiser, H. K., Wiech, G. & Quinkert, G., *Fortschr. Chem. Forsch.*, 1974, **46,**, 181.
12. Caldwell, R. A., Sakuragi, H. & Majima, T., *J. Am. Chem. Soc.*, 1984, **106**, 2471.
13. Schuster, D. I., in *Rearrangement in Ground and Excited States*, ed. de Mayo, P., Academic, 1980, Vol. 3, p. 232.
14. Houk, K. N., *Chem. Rev.*, 1976, **76**, 1.
15. Dauben, W. G., Lodder, G. & Ipaktschi, J., *Top. Current Chem.*, 1975, **54**, 73.
16. Woodward, R. B. & Hoffmann, R., *The Conservation of Orbital Symmetry*, Verlag Chemie/Academic, 1970.
17. Sadler, D. E., Wendler, J., Olbrich, G. & Schaffner, K., *J. Am. Chem. Soc.*, 1984, **106**, 2064.
18. Schaffner, K. & Demuth, M., in *Rearrangements in Ground and Excited States*, ed. de Mayo, P., Academic, 1980, Vol. 3, p. 281.
19. Schuster, D. I., *Acc. Chem. Res.*, 1978, **11**, 65.

Cycloadditions are of considerable importance in synthetic organic chemistry since they provide access to cyclic compounds from acyclic molecules. Of these reactions the ground state addition of a diene to an alkene, the Diels–Alder reaction, is the best known and results in the simultaneous formation of two carbon–carbon σ-bonds. This is shown for reaction of the s-*cis* or cisoid conformation of butadiene with ethene.

The application of the principle of conservation of orbital symmetry, by Woodward & Hoffmann,[1] to cycloaddition and other pericyclic processes in the mid-sixties stimulated considerable research into such reactions, pointed to hitherto unthought of possibilities and has provided a major contribution to our understanding of organic chemistry. These ideas stimulated detailed studies directed towards elucidating the mechanism of cycloadditions.[2] At the present time the precise mechanistic details of many such reactions remain a matter of debate; cycloadditions proceed by a variety of distinguishable routes, not all are concerted and each can exhibit subtleties inviting a study of details at first not apparent.

4.1   **The $[_\pi 2 + _\pi 2]$ cycloaddition reaction of alkenes**[1-10]

The cycloaddition of two alkene molecules to give a cyclobutane is a well-known photochemical reaction. The reverse process, the fragmentation of a cyclobutane to two alkenes, is an exothermic process under thermal reaction conditions releasing the strain energy of the four-membered ring. In the cycloaddition the energy required to form the strained cyclobutane moiety is provided by the absorbed radiation. The stability of the product to the reaction conditions results from the absence of an absorbing chromophore in the product; since the product does not absorb light it does not undergo carbon–carbon bond rupture under the irradiating conditions under which it is formed.

The addition of two ethene molecules with maximum symmetry of approach requires that they lie on top of each other. The cycloaddition can then proceed in a suprafacial manner to each component and the reaction may be represented formally as a $[_{\pi}2_s +_{\pi}2_s]$ process. Such a light induced process is orbital symmetry allowed. The number of $(4q+2)_s$ and $(4r)_a$ components is even because each of the $_{\pi}2_s$ components is of the type $(4q+2)_s$ where the integer q is zero and there are no $(4r)_a$ components. This

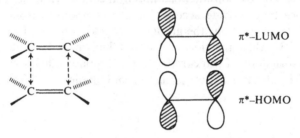

$[_{\pi}2_s +_{\pi}2_s]$ cycloaddition

reaction is therefore symmetry allowed in the excited state. Consideration of the frontier orbitals[4,5] involved in this reaction shows the HOMO (highest occupied molecular orbital) of the excited alkene is the π^*-orbital and the LUMO (lowest unoccupied molecular orbital) of the ground state alkene to which it adds is also the π^*-orbital. The interaction of these two orbitals is favourable since the phases at the reacting carbons are of the same sign. This interaction represents a favourable contribution to the transition structure for the reaction and is analogous to stating that the reaction is symmetry allowed.[6]

Electrocyclic reactions which are discussed in section 2.2 are a special case of cycloadditions. For example, the conversion of butadiene to cyclobutene can be regarded as a $[_{\pi}2+_{\pi}2]$ reaction.

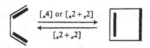

The $[_{\pi}2+_{\pi}2]$ reaction of alkenes can occur intramolecularly whereupon bicyclic molecules are obtained and this is shown for the reaction of diene (1). The same diene can undergo a crossed $[_{\pi}2+_{\pi}2]$ cycloaddition (signified

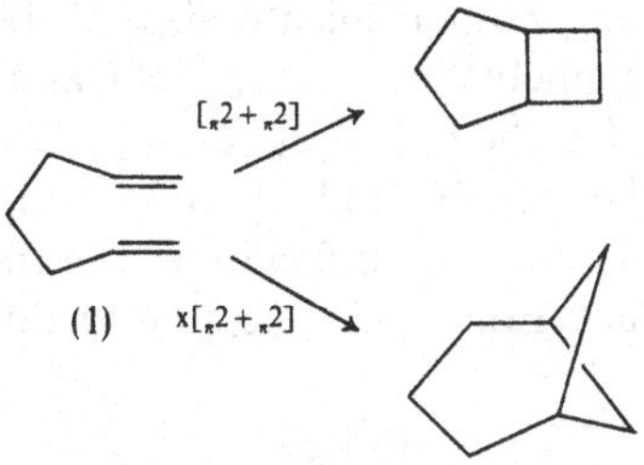

(1)

as $x[_\pi 2 +_\pi 2]$ to give a bicyclo[3,1,1]heptane (see section 2.4).

An alternative geometry of reaction of two ethene molecules occurs from approach of the alkenes in orthogonal planes. For an appropriately

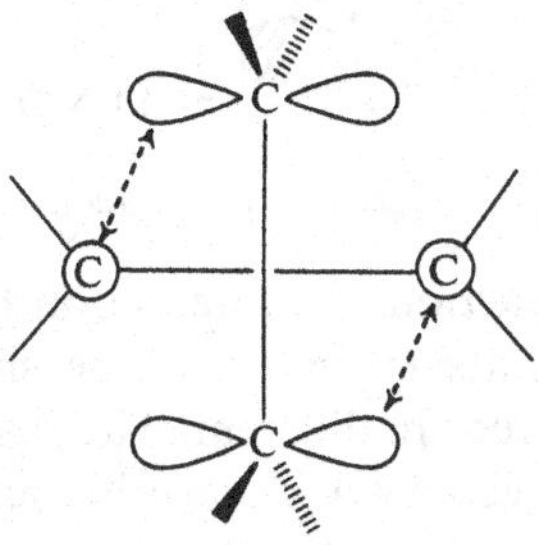

$[_\pi 2_s +_\pi 2_a]$ cycloaddition

substituted alkene this mode of cycloaddition leads to a product with a different stereochemical relationship of the substituents than from the reaction where the two alkenes approach with maximum symmetry (scheme 4.1). This alternative mode of reaction requires the addition of one

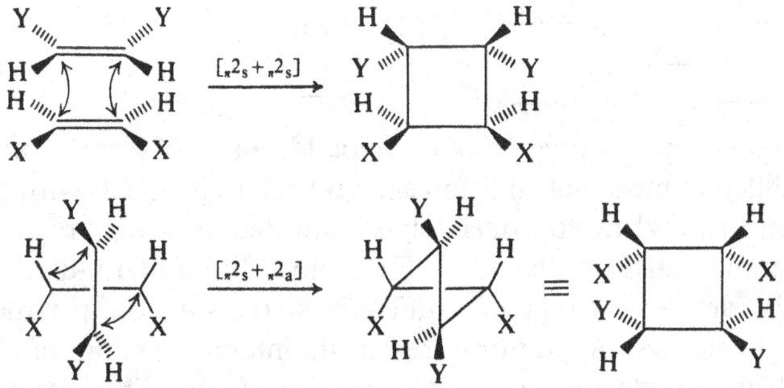

Scheme 4.1

$_\pi 2$ component in a suprafacial manner to the other in an antarafacial manner. Thus the cycloaddition can be described as a $[_\pi 2_s +_\pi 2_a]$ process where the molecule drawn vertically undergoes *antarafacial addition* and that drawn horizontally undergoes *suprafacial addition*. This mode of reaction is allowed thermally, but not photochemically since there is only one component of the type $(4q + 2)_s$, namely the $_\pi 2_s$ component, and none of the type $(4r)_a$. The frontier orbitals for the $[_\pi 2_s + _\pi 2_a]$ cycloaddition only have favourable interaction when the HOMO and LUMO of the ground state are considered. This favourable interaction facilitates the transition between starting alkenes and product. For a $[_\pi 2_s + _\pi 2_a]$ cycloaddition to

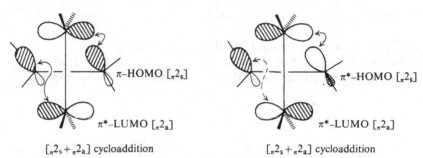

$[_\pi 2_s + _\pi 2_a]$ cycloaddition $[_\pi 2_s + _\pi 2_a]$ cycloaddition

occur from a photochemically induced reaction, the interaction between the HOMO (π^*) and LUMO (π^*) orbitals is unfavourable since the phases of the orbitals at the reacting centres are different. The reaction is therefore not expected to occur with this geometry under photochemical conditions. It is probable that the transition state structure for the reverse reaction, the thermal rupture of cyclobutane, will exhibit considerable twisting as the molecule attempts to adopt a geometry which will allow for a concerted symmetry-allowed cleavage. If the geometrical constraints are too great to attain this geometry, then the reaction may occur in a non-synchronous manner via a diradical intermediate which subsequently undergoes fragmentation.

The stereochemical outcome of a cycloaddition is of prime importance since the different modes of addition can give rise to products with differing stereochemistries when appropriately substituted addends are employed. This is demonstrated in scheme 4.1 for cycloaddition of 1,2-disubstituted alkenes. In addition to giving different stereoisomeric products, the differing geometries of addition result in interconversion of different orbitals between starting reactants and products. This results in a particular process being orbital symmetry forbidden and therefore unlikely to occur, or orbital symmetry allowed and likely, in the absence of overriding steric factors, to be observed.

For the maximum symmetry of approach of two alkene molecules (fig. 4.1) a correlation diagram can be constructed which shows how the orbitals of the starting alkenes are transformed into those of the product. As discussed for electrocyclic reactions (section 2.2), only those orbitals that are most involved in the reaction need be considered even though it must be realised that all the molecular orbitals will undergo some energy change as the bond lengths and bond angles change from substrate(s) to product(s). The energy requirement for these processes will contribute to the activation energy for the reaction. Nonetheless, it is the higher occupied molecular orbitals and the lower unoccupied molecular orbitals that are particularly important in facilitating reaction or in dictating a high energy barrier to the reaction. Indeed it can be shown that the stereochemistry of a number of molecular processes is governed by the symmetry of the HOMO (see, for example, sections 2.2, 2.4).

To construct a correlation diagram the energy levels of the π- and π^*-molecular orbitals of the reactant ethene molecules are placed on one side of a diagram and the σ- and σ^*-orbitals of cyclobutane on the other (fig. 4.2). These orbitals are then classified with respect to the symmetry elements of the molecular transformation which for this reaction are three planes of symmetry (1, 2 and 3) (fig. 4.1). In the ground state reaction the direction of approach leads to the $\pi_1 + \pi_2$ and $\pi_1 - \pi_2$ orbital combinations of the two ethene molecules. These orbitals, drawn as the cross section in plane 3 (fig.

Fig. 4.1. Maximum symmetry approach of two ethene molecules.

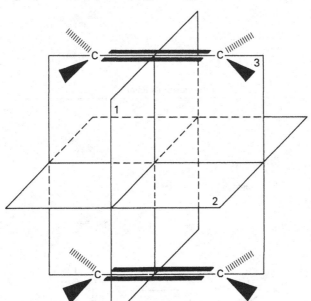

4.3) at finite distances of approach of the two π-systems, can be seen to be of different energy, $\pi_1 + \pi_2$ being lower than $\pi_1 - \pi_2$. The orbitals are of course symmetrical (S) with respect to plane 3 and are respectively symmetrical–symmetrical (SS) and symmetrical–antisymmetrical (SA) with respect to planes 1 and 2. A similar consideration of the antibonding orbitals of the two ethenes at finite distance leads to the $\pi_1^* + \pi_2^*$ and $\pi_1^* - \pi_2^*$ orbitals, and these are classified as AS and AA respectively with reference to planes 1 and 2 (fig. 4.3). The orbitals of the product, cyclobutane can now be considered in an analogous manner by taking the combinations $\sigma_1 + \sigma_2$, $\sigma_1 - \sigma_2, \sigma_1^* + \sigma_2^*$ and $\sigma_1^* - \sigma_2^*$ with respect to the symmetry planes 1 and 2 the combinations are respectively, SS, AS, SA and AA (fig. 4.4). The correlation diagram for the reaction of two ethenes to give cyclobutane is now constructed (fig. 4.5). As the two ethene molecules approach one another the SS and AS orbitals, which are symmetrical with respect to plane 2, are stabilised by the interaction. In like manner, the SA and AA orbitals are destabilised. Conversely, as the cyclobutane is being pulled apart to give

Fig. 4.2. Energy levels of orbitals involved in the formation of cyclobutane from two ethene molecules.

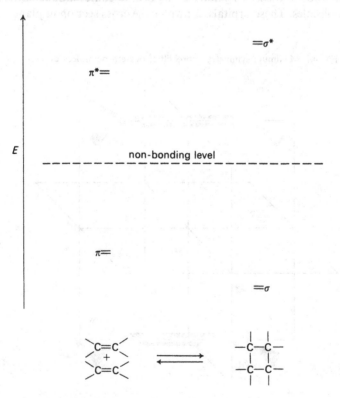

two ethene molecules, the SS and AS orbitals are bonding and resist the motion. On the other hand the motion is favourable for the SA and the AA orbitals. The correlation diagram is completed by joining orbitals of like symmetry bearing in mind the quantum mechanical non-crossing rule (section 2.2).

In the ground state reaction, the ethene molecules are predicted to combine with the electrons in the SS and SA π-orbitals to give, with

Fig. 4.3. Energy levels of two ethene π-orbitals at finite distance.

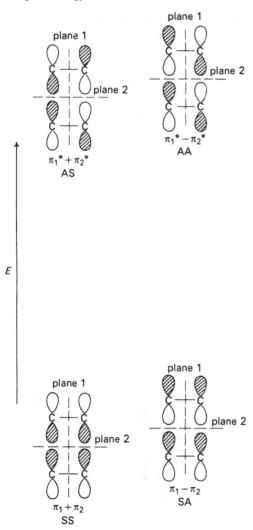

symmetry conservation, cyclobutane with two electrons in the bonding SS σ-orbital, and two electrons in the anti-bonding SA σ^*-orbital (fig. 4.5). This is clearly an unfavourable high-energy process, and the magnitude of the energy barrier would be approximately that required to raise two bonding electrons to the nonbonding level ($c.$ 481 kJ mol^{-1}). Similarly, for cyclobutane to rupture to two ethene molecules in a concerted manner, the ethenes would be produced with two electrons in the antibonding AS orbital. The thermal ground state $[_\pi 2_s + _\pi 2_s]$ cycloaddition is thus symmetry forbidden. However, if one of the ethene molecules, or the

Fig. 4.4. Energy levels of cyclobutane delocalised σ-orbitals.

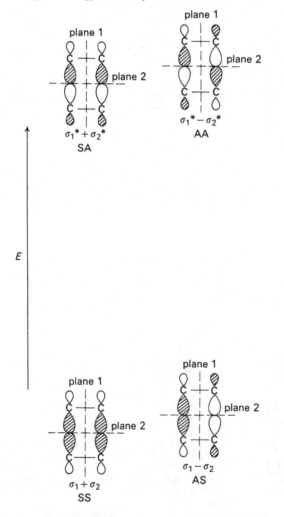

cyclobutane, is in the first excited state there is no such symmetry-imposed barrier to the addition, or cycloreversion, since the starting material with one electron in the excited state gives a product with one electron in the excited state: $(SS^2, SA^1, AS^1 \rightarrow SS^2, AS^1, SA^1)$ (fig. 4.6). The photochemical $\left[_\pi 2_s +_\pi 2_s\right]$ cycloaddition reaction is thus symmetry allowed.

An electronic state, correlation diagram for the formation of cyclobutane from two molecules of ethene (fig. 4.7) further illustrates the principle of conservation of orbital symmetry. The ground state of the two ethenes SS^2, SA^2 correlates with the high-energy, doubly excited state of cyclobutane SS^2, SA^2. The converse, the ground state of cyclobutane correlating with the doubly excited state of the two ethenes, also applies. However, levels of like symmetry are not allowed to cross (the quantum mechanical non-crossing rule) and electron interaction will force a correlation between the two ground states. Nevertheless, the reaction will have a high activation enthalpy resulting from the intended, but avoided, crossing. This is estimated as *c.* 481 kJ mol^{-1} the energy required to afford a doubly excited state. As discussed above, the lowest excited state of the ethenes generated by photochemical excitation (SS^2, SA^1, AS^1) correlates directly with the first excited state of cyclobutane and thus the $\left[_\pi 2_s +_\pi 2_s\right]$ cycloaddition of alkenes is a symmetry-allowed photochemical reaction. However, it must be emphasised that photochemical reactions can be

Fig. 4.5. Correlation diagram for the formation of cyclobutane from two molecules of ethene.

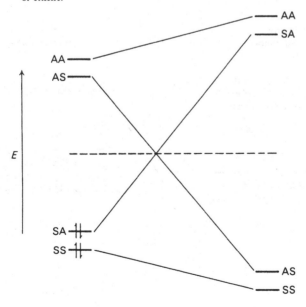

Fig. 4.6. The excited state $[_\pi 2_s + _\pi 2_s]$ cycloaddition of two molecules of ethene.

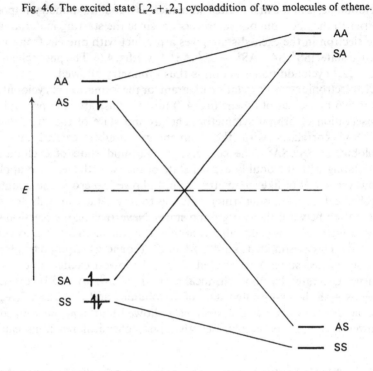

Fig. 4.7. Electronic state diagram for formation of cyclobutane from two molecules of ethene.

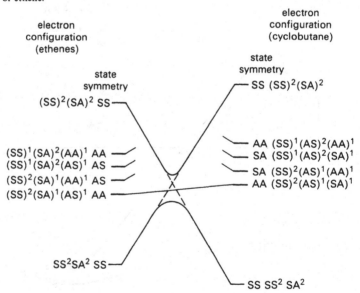

complicated by the possibility of different excited states (and different vibrational levels within these excited states) being involved in the reaction. It is often difficult to identify the excited state and the vibrational level from which reaction occurs.

The alternative $[_\pi 2_s + _\pi 2_a]$ ground state cycloaddition course was discussed earlier. It does not involve the symmetrical least-motion path and similarly the reverse process, the cycloreversion of cyclobutane to two ethene molecules ($[_\sigma 2_s + _\sigma 2_a]$), likewise does not involve the least-motion pathway. These processes are symmetry allowed as ground state reactions,

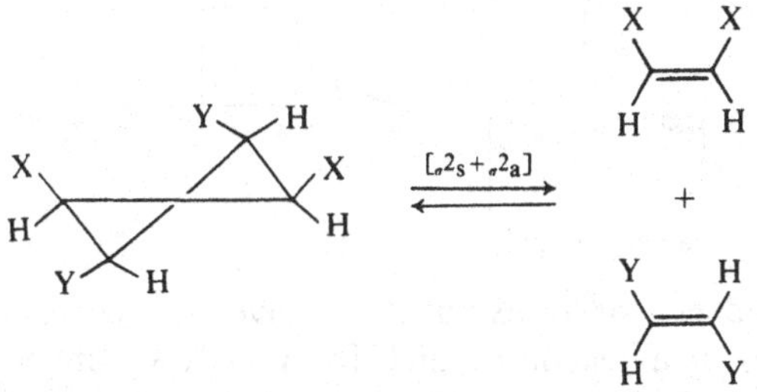

but require high-energy steric interactions in the transition state structure and will only be favoured when the system(s) are twisted by constraints within a molecule. The formation of substituted cyclobutanes from twisted alkenes (scheme 4.2) may be examples of $[_\pi 2_s + _\pi 2_a]$ cycloaddition, although

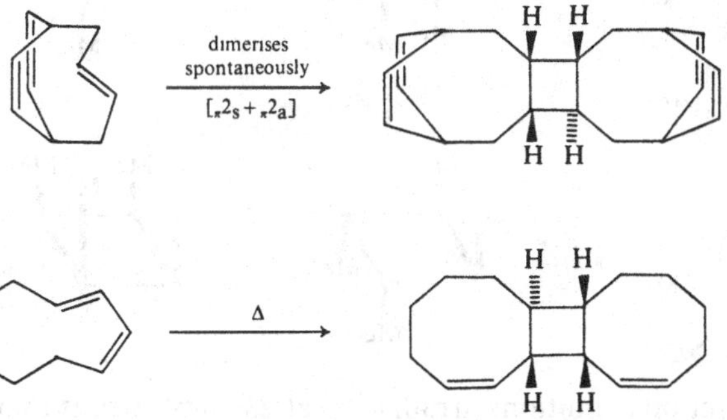

Scheme 4.2

in the latter case the formation of two other isomeric cyclobutanes has led to the suggestion that the dimerisation is not a concerted reaction. In the absence of steric constraints a ground state $[_\pi 2 + _\pi 2]$ cycloaddition or the corresponding cycloreversion reaction is unlikely to show a preference for this symmetry-allowed process.

A large number of $[_\pi 2 + _\pi 2]$ photocycloaddition reactions are known.[7-10] However, there are relatively few examples of photodimerisation of simple non-conjugated acyclic alkenes compared with those known for conjugated and small rings alkenes. In general, $[_\pi2+_\pi2]$ cycloadditions proceed from the triplet excited state, but since a spin-pairing process is necessary for bond formation in a triplet reaction the process is precluded from being concerted and dipolar and diradical intermediates have been postulated for these reactions.

radical ion pair

Irradiation of liquid *cis*- or *trans*-but-2-ene gives the corresponding isomeric cyclobutanes in a reaction which for very low conversions is stereospecific. The more facile *cis–trans* interconversion of the but-2-enes

under the reaction conditions (irradiation at 230 nm), means that at longer reaction times mixtures of all three isomeric cyclobutanes are formed. The stereospecificity observed at low conversions can be accounted for if the dimerisation reaction is concerted with the singlet π–π^* excited state adding to a ground state molecule, i.e. a $[_\pi2_s+_\pi2_s]$ cycloaddition. However, the formation of a singlet exciplex by interaction of singlet excited state–ground state molecules is possible. Collapse of this complex either to a cycloadduct or a diradical (which undergoes ring closure at a rate more rapid than bond rotation and loss of stereochemistry) also explains the observed product. The $[_\pi2+_\pi2]$ cycloadditions of triplet excited state non-

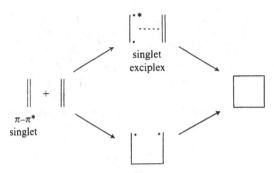

diradical intermediate

conjugated acyclic alkenes are not common, and it is thought that the triplets of these species undergo deactivation sufficiently rapidly that addition processes are not competitive. On the other hand, many cyclic alkenes undergo triplet sensitised $[_\pi 2 + _\pi 2]$ reactions. The sensitiser must be chosen with care since only a few sensitisers have sufficient energy (c. 310–335 kJ mol^{-1}) to effect sensitisation of the alkene. For example norbornene, dimerises when acetophenone ($E_T = 310$ kJ mol^{-1}) is used as the sensitiser but not when benzophenone ($E_T = 285$ kJ mol^{-1}) is used since the latter has insufficient energy to excite norbornene to the triplet excited state. Triphenylcyclopropene undergoes sensitised dimerisation as does cyclopentene.

The photodimerisation of alkenes in solution frequently leads to isomeric products while direct photolysis on the solid phase is frequently stereospecific. For example, maleic anhydride yields the isomeric *cis-cisoid-cis* and *cis*-transoid-*cis* dimers (2) and (3) on solution-phase photolysis, whereas only the former is obtained from photolysis of the solid material. For *trans*-cinnamic acid there are three crystalline modifications, α, β, and γ. In the α- and β-forms the molecules lie 'head-to-tail' and 'head-to-head' respectively and the non-bonded carbon–carbon distances are similar, c. 0.36 nm. Solid state photolysis of these two forms leads to the

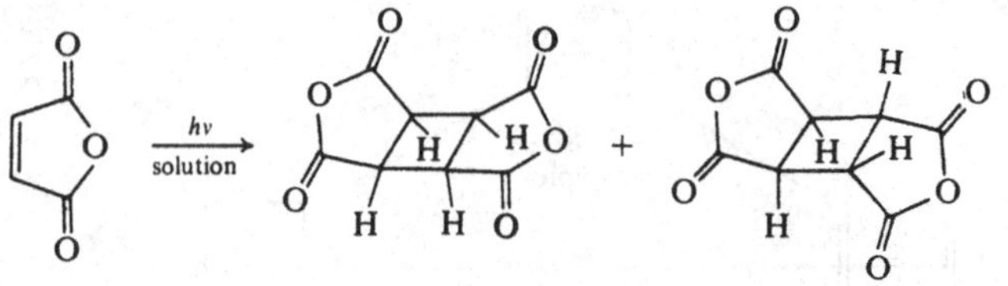

cis-cisoid-*cis*   cis-transoid-*cis*
(2)   (3)

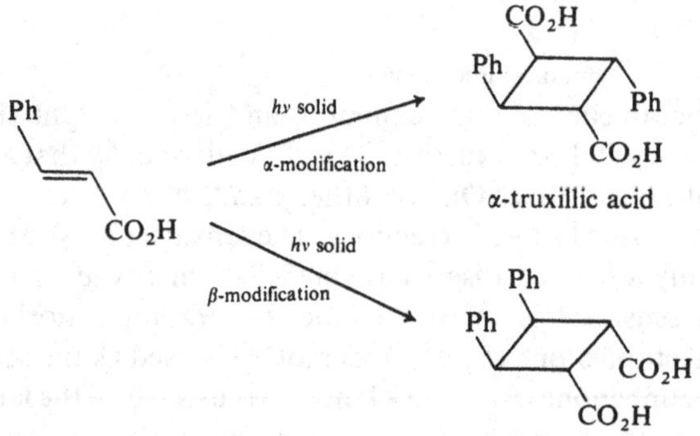

α-truxillic acid

β-truxillic acid

specific α- and β-truxillic acids respectively. The third crystalline form, the γ-form, is photochemically inactive because of the greater separation of the molecules (*c.* 0.5 nm) in the crystal.

A variety of formally non-conjugated dienes undergo internal cycloaddition on photolysis. For example, norbornadiene, a molecule which possesses no formally conjugated chromophore, gives quadricyclene on either direct or sensitised irradiation in solution. Norbornadiene

norbornadiene   quadricyclene

absorbs electromagnetic radiation at *c.* 210 nm while a normal, isolated, alkenic bond does not absorb radiation above 195 nm. The shift of the absorption maximum to longer wavelength results from the molecular geometry imposing an interaction between the π-orbitals on the underface of the molecule which clearly facilitates the cycloaddition. The norbornadiene–quadricyclene reaction is reversible and the opening of the strained cyclopropane rings occurs on sensitised photolysis (or with a transition metal catalyst) with consequent relief of strain. This is analogous to the cleavage of a carbon–carbon bond in the *cis–trans* isomerisation of cyclopropane derivatives. While norbornadiene is a 1,4-diene, the di-π-methane rearrangement (section 2.3) is suppressed in favour of cycloaddition. On the other hand, barrelene does not undergo cycloaddition but gives a di-π-methane rearrangement product.

barrelene

Replacement of the methylene bridge of norbornadiene by a —CH=CH— moiety effects a change in the stereochemical relationship of the π-bonds which is sufficient to alter the course of the reaction.

Internal cycloadditions generally yield highly strained products which are unobtainable by other routes, as illustrated by the formation of prismane from 'Dewar' benzene. A successful synthesis of cubane involved an internal $[_{\pi}2+_{\pi}2]$ cycloaddition and this type of reaction is not confined to carbocyclic molecules (e.g. (4) to (5)). Intramolecular photocycloadditions such as this are currently receiving considerable attention for

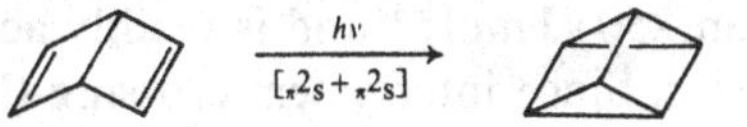

prismane

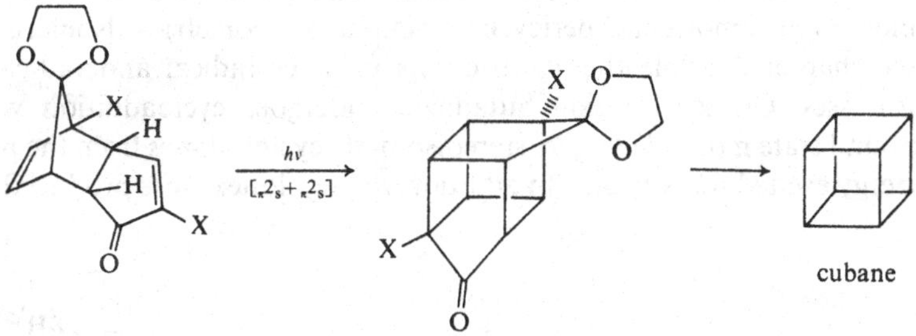

cubane

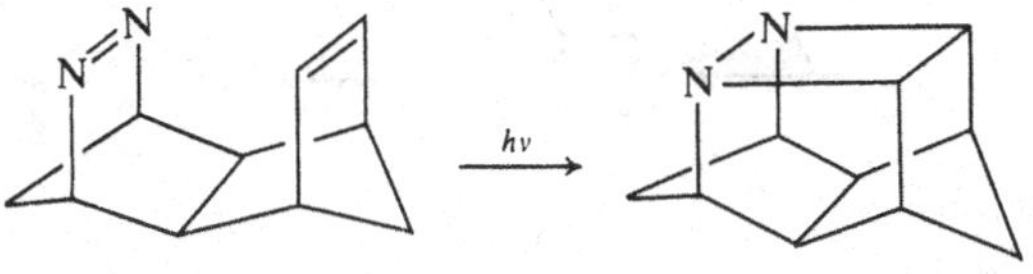

(4)   (5)

their ability to 'trap' sunlight. Catalytic and thermal cleavage regenerates starting material with release of energy and thus 'cage' compounds can be regarded as energy stores.

The intramolecular [2+2] reaction is not confined to the addition of two π-bonds. The addition of a π-bond to a strained cyclopropyl σ-bond is well known and is classified as a $[_\pi 2 + _\sigma 2]$ cycloaddition. The reaction requires the three-membered ring to be orientated to allow overlap of the edge of the cyclopropyl group with the orbitals of the alkene. An example occurs for photolysis of *exo*-tricyclo$[3,2,1,0^{2,4}]$octene (6) and a similar reaction occurs for the dicyclopropyl compound (7).

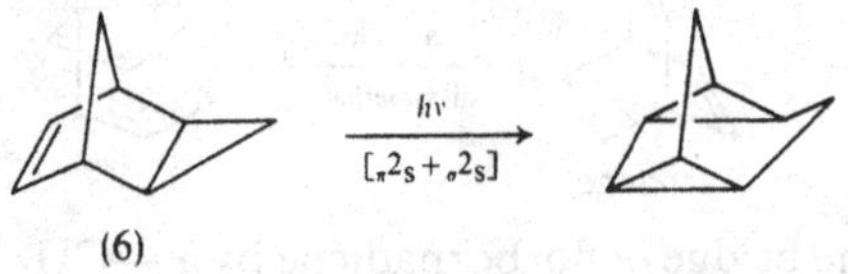

(6)

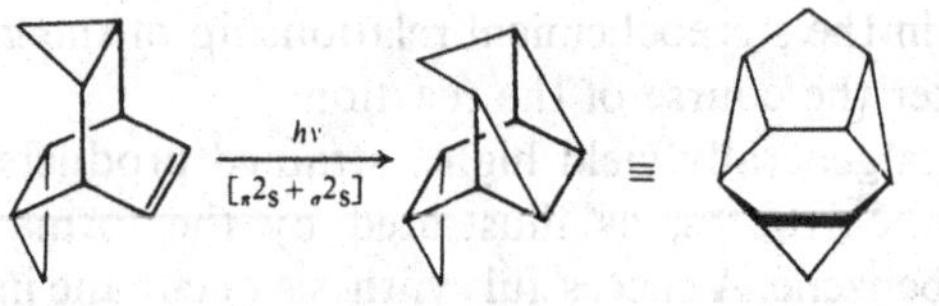

(7)

The photochemistry of 1,3-dienes is well established (see section 2.2) as the molecules absorb above 215 nm. Furthermore, the energy of the triplet excited state is generally less than $250 \, \text{kJ} \, \text{mol}^{-1}$ and is readily accessible using a variety of triplet sensitisers. Since intersystem crossing (ISC) for dienes is inefficient, direct irradiation produces the singlet excited state and the chemistry of this state can be studied readily. In general the S_1 reactions undergo intramolecular pericyclic reactions or geometrical isomerisation (see chapter 2) while the T_1 reactions involves radical and radical-like processes. On sensitisation butadiene undergoes cycloaddition with a ground state molecule to give stereoisomeric cyclobutanes from the higher energy excited *trans*-triplet. In addition vinylcyclohexene is produced from

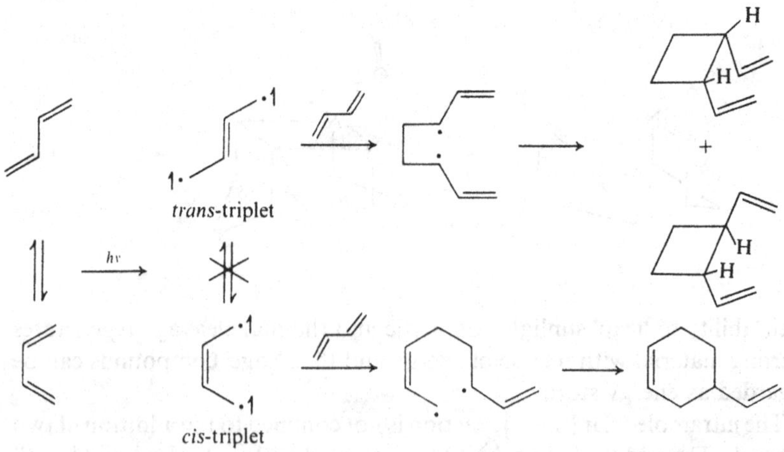

Scheme 4.3

the lower energy excited *cis*-triplet by a non-concerted pathway (scheme 4.3). Interconversion of the *cis* and *trans* excited states is considered to be slow since population of the butadiene ψ_3-orbital increases the double-bond character of the C_2–C_3 bond.

4.2   [2 + 2] Cycloaddition of carbonyl compounds to alkenes – oxetane formation[1–3,10,11]

One of the first photocycloaddition reactions to be studied was the formation of oxetanes from the addition of carbonyl compounds to alkenes – a reaction known as the Paterno–Büchi reaction – and formally represented as a $[_\pi 2 + _\pi 2]$ cycloaddition process.

A concerted mechanism[10] for the reaction of the n–π* state of a carbonyl compound with an alkene is not expected because of the difficulty in obtaining appropriate orbital overlap in the transition state structure for the reaction. The n- and π*-orbitals of the carbonyl group are orthogonal and overlap of these with the π/π*-orbital of the alkene is geometrically difficult to achieve due to steric requirements. The Paterno–Büchi reaction is therefore a diradical process and proceeds by attack of a singlet or triplet n–π* state of the carbonyl compound on the alkene.

The question arises as to whether it is the n-orbital of the n–π* excited state of the ketone which reacts first with the π-orbital of the alkene or the π*-orbital which reacts with the π*-orbital of the alkene. For the n-orbital

n → π*

perpendicular approach

of the carbonyl to attack the alkene with oxygen–carbon bond formation as the first step in the reaction the substrates would of necessity be perpendicular and the radical orbitals of the diradical so produced would be orthogonal. For such a process to occur the alkene would have to have electron donating ability since an electron on the π-orbital would be transferred to the n-orbital on oxygen. Such a process would be expected when carbonyl compounds undergo reaction with electron-donating or electron-rich alkenes and for such a reaction the excited state of the carbonyl compound can be regarded as an electrophile.

For the π^*-orbital radical of the excited carbonyl to attack the alkene the substrates would be required to be in a parallel configuration. Such a

parallel approach

reaction would involve the transfer of an electron from the π^*-orbital of the carbonyl to the vacant π^*-orbital of the alkene with concomitant formation of a carbon–carbon bond and a tritopic† diradical, the oxygen radical centre being essentially degenerate. The formation of the carbon–carbon bond from this parallel geometry is symmetry allowed. The increase of topicity in the reaction from a ditopic starting material to a tritopic intermediate generates an allowed surface pathway for the parallel approach. This process is likely to occur when the alkene is electron deficient and it can be regarded as nucleophilic attack of the excited carbonyl compound on the alkene. The alternative process, involving carbon–oxygen bond formation from the parallel arrangement of alkene and carbonyl, is symmetry forbidden generating a ditopic diradical, and can therefore be ignored.

The formation of diradical intermediates by either carbon–oxygen or carbon–carbon bond formation between the carbonyl group and the alkene will result in some selectivity as to the carbon of the alkene to which bonding occurs. Steric and electronic factors are undoubtedly important, but in general one would expect to find the alkene bonding in such a way as

† Topicity[10,11] is defined as the total number, and nature, of available sites generated in the primary reaction process.

to generate the more substituted, and hence more stable, radical centres of the possible diradicals. For example, reaction of benzophenone with 2-methylpropene gives a 9:1 mixture of oxetanes with the dominant isomer formed from the more stable diradical, *viz.* oxygen–carbon bonding must occur in the first step of the reaction.

Additional evidence for the presence of diradical intermediates in these reactions stems from the observation of products of radical

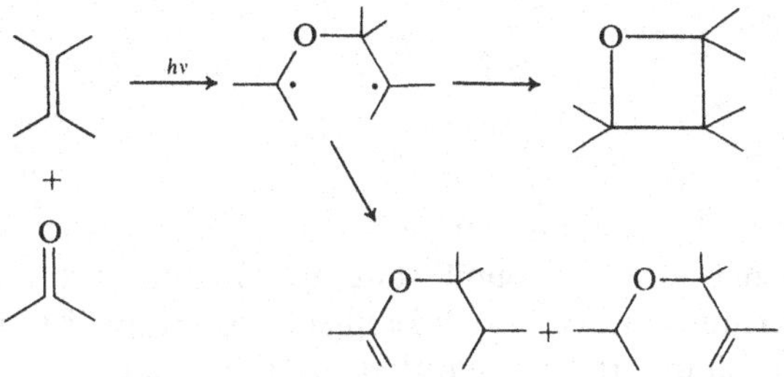

disproportionation. Moreover, the intermediacy of diradicals follows from the photochemical addition of benzophenone to the vinylcyclopropane (8). In addition to normal oxetane formation to give (9), rearrangement of the

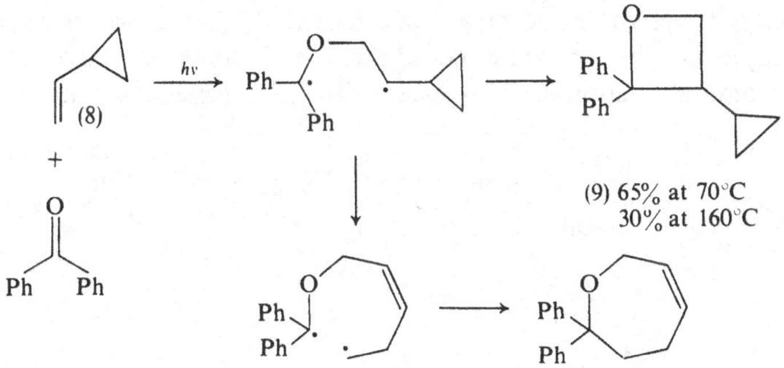

(9) 65% at 70°C
30% at 160°C

(10) 15% at 70°C
65% at 160°C

cyclopropyl carbinyl diradical intermediate occurs to give the ring-expanded product (10). As the temperature is increased the reaction partitions to favour the rearrangement product.

The attack of the excited carbonyl on the alkene to form a diradical intermediate is generally thought to be preceded by the formation of an exciplex or a radical ion pair. Exciplexes have been proposed as intermediates to account for the inefficiency of the reaction. In some cases the n–π* state is completely quenched by the alkene, but more generally it leads to products with less than 100% efficiency. It is for this reason a deactivation channel to the ground state via an exciplex has been postulated. Furthermore, rate constants for quenching are much faster than those expected for simple radical addition reactions and argue against direct formation of the diradical. In some addition reactions the product structures are incompatible with the intervention of the most stable diradical and this again suggests that there is a precursor to the diradical. The results do not distinguish between this precursor being a radical ion pair or an activated complex and the term exciplex is sometimes used rather loosely. It is possible that an exciplex has strong charge-transfer character. A simple diradical pathway should give almost solely the 2-alkoxyoxetane (13) from the reaction of acetone with vinyl ether (scheme 4.4) if the stability

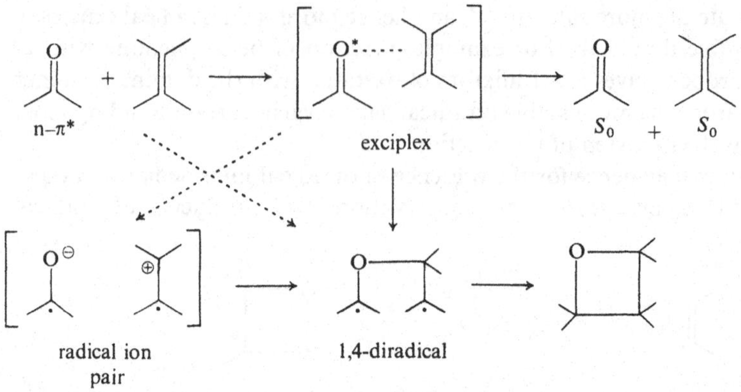

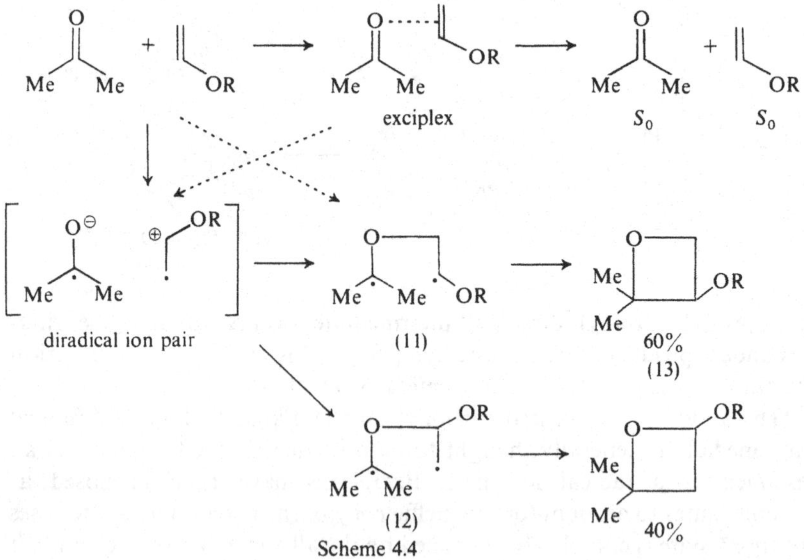

Scheme 4.4

of the possible diradical intermediates is dominant in determining the mechanism; diradical (11) would be favoured over diradical (12). In fact the reaction shows little regioselectivity and it has been suggested that this is consistent with a diradical ion pair preceding diradical formation. The collapse of the ion pair to the diradical is considered not as regioselective as direct addition of an n–π* state to the enol ether (scheme 4.4).

If a diradical is produced by attack of the carbonyl in a perpendicular configuration to the alkene then, depending on the excited state of the carbonyl (singlet or triplet), either a singlet or a triplet exciplex or radical ion pair will be formed. If the diradical is in the singlet excited state some spin attraction will exist between the termini of the diradical. If however the diradical is formed from the triplet state of the ketone then the radical centres at the termini of the diradical will exhibit spin repulsion. It is not surprising therefore that the different excited states of a given substrate will partition differently to products.

When benzaldehyde is irradiated in the presence of 2-methylbut-2-ene the four possible stereoisomeric oxetanes are formed (scheme 4.5). It is

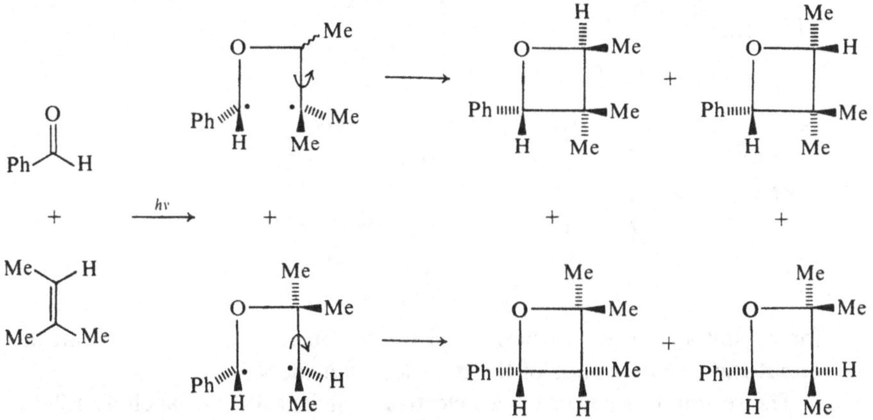

Scheme 4.5

reasonable to suppose that the reaction involves the triplet state and is not concerted. For benzaldehyde ISC is too fast ($> 10^{10}$ s^{-1}) from the S_1 state for reaction from this state to be competitive. The diradical intermediates produced will therefore be triplet species and spin inversion, a process with a definite time requirement, must precede ring closure. Moreover, the diradicals have sufficient lifetime to allow rotation to occur about the terminal bonds before ring closure. After ISC occurs closure to oxetane or fragmentation is rapid since there is no no spin imposed barrier to either reaction process.

If the alkene possesses a lower excitation energy than the carbonyl, energy transfer (or sensitisation), to produce an electronically excited state of the alkene can occur. Electron-rich ethenes are sensitised by a π–n interaction while for electron deficient ethenes π^*–π^* interaction dominates. The rate constant for π–n sensitisation is the larger of the two.

For appropriately substituted alkenes, geometrical isomerism is often observed. This can occur by sensitisation of the alkene or by collapse of a sufficiently long-lived 1,4-diradical intermediate which has undergone rotation about the terminal bonds.

In general there is more stereospecificity in oxetane formation when the reaction occurs from the S_1 than from the T_1 excited state and there is less geometrical isomerisation from S_1 than from T_1. Acetone reacts with the alkene (14) to give both *cis*- and *trans*-oxetane and this has been accounted

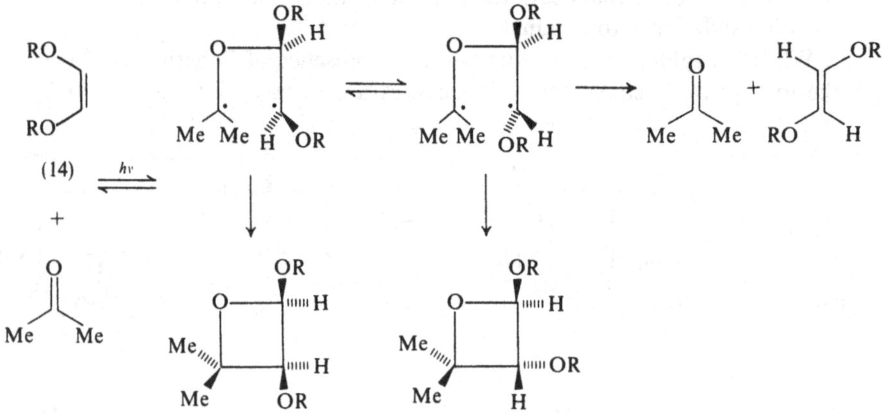

Scheme 4.6

for by initial oxygen–carbon bond formation and subsequent rotation about the termini bonds of the diradical (scheme 4.6).

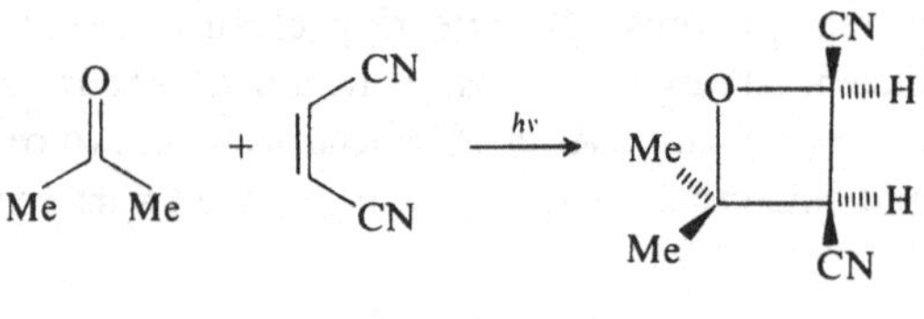

The reaction of ketones with electron-deficient alkenes such as 1,2-dicyanoethene is somewhat different from the reaction with electron-rich alkenes. Only the singlet n–π^* state of the ketone gives oxetanes. Quenching of this state does not effect geometrical isomerism which is a side-reaction. Oxetane formation from the singlet state is stereospecific as shown for addition of acetone to 1,2-dicyanoethene, but the reaction is inefficient. The reaction is thought to occur through the parallel approach of the ketone and the alkene, carbon–carbon bond formation, and rapid closure of the spin-paired diradical intermediate thus produced. The inefficiency must result from the reversible collapse of the reacting intermediate species.

The regiospecificity of the reaction between 2-cyanopropene and singlet acetone reflects the stability of the diradicals possible from carbon–carbon bond formation. On triplet sensitisation the reactions with electron-deficient alkenes leads to *cis–trans* isomerisation or to dimerisation of the

not observed

alkene. The failure to observe oxetane formation from the triplet state, when the reactants interact in a parallel configuration, must be due to fragmentation being more rapid than spin inversion and oxetane formation.

4.3 **$[_{\pi}2+_{\pi}2]$ Cycloaddition reactions of αβ-unsaturated carbonyl compounds**[1–3,10,12,13]

The photochemical dimerisation of αβ-unsaturated ketones provides a further class of $[_{\pi}2+_{\pi}2]$ reactions[1–3] and is demonstrated for 3-methylcyclopent-2-enone and 3-methylcyclohex-2-enone with the formation of head-to-head dimers and head-to-tail dimers. In general there is a preference for *cis*-transoid-*cis* head-to-tail dimerisation. However, when 3-methylcyclopentenone is dissolved in micelles and photolysed an almost quantitative yield of head-to-head dimer is formed. The reactions occur[10] with light of wavelength greater than 300 nm and therefore n–π* excitation is involved. ISC is efficient for enones and the reactions are believed to occur from a triplet manifold. However, this may be the n–π* and/or the π–π* triplet depending upon the particular substrate and the precise conditions employed. Irradiation of cyclopentenone in the presence of penta-1,3-diene, under conditions where only the enone absorbs, causes geometrical isomerisation of the diene as well as dimerisation of the enone. The dimerisation is thought to proceed by attack of an n–π* triplet, or in

head-to-head   head-to-tail   head-to-tail

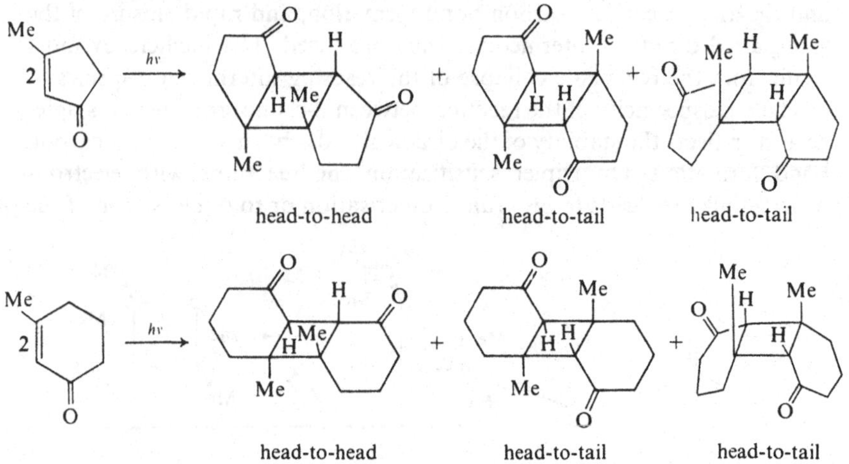

head-to-head   head-to-tail   head-to-tail

some cases a π–π* triplet, excited species on a ground state enone molecule. Whilst the ratio of head-to-head to head-to-tail dimerisation is dependent on the solvent and the concentration of substrate it is unaffected by the method of triplet population (ISC or sensitisation).

$\alpha\beta$-Unsaturated ketones also react[12,13] with cyclic and acyclic alkenes and the reactions although usually intermolecular can be intramolecular with substrates of appropriate structure, and selected examples are shown in schemes 4.7 and 4.8. The reaction of cyclohexenone with ethene gives the bicyclic addition product with *cis*-stereochemistry and the analogous reaction with cyclopentene gives the *cis-transoid-cis* adduct as the major product (scheme 4.7). These reactions find a useful place in synthesis since

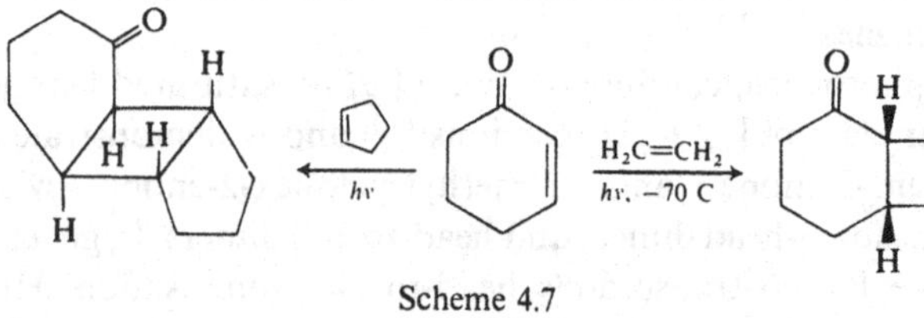

Scheme 4.7

two σ-bonds are formed to give a strained cyclobutane which is often difficult to obtain from other processes. The intramolecular reaction of the acetoxycyclohexenone (scheme 4.8) gives a mixture of two stereoisomers and demonstrates that the cycloaddition is effective even with an acetoxy substituent on the enone. This extends the use of the reaction in synthesis since the ester products can be hydrolysed and the β-hydroxyketones with base undergo a reverse aldol reaction. The photolysis of carvone is a reaction that has been known for a long time and although the yield of intramolecular adduct is *c.* 50%, the reaction has only a modest quantum

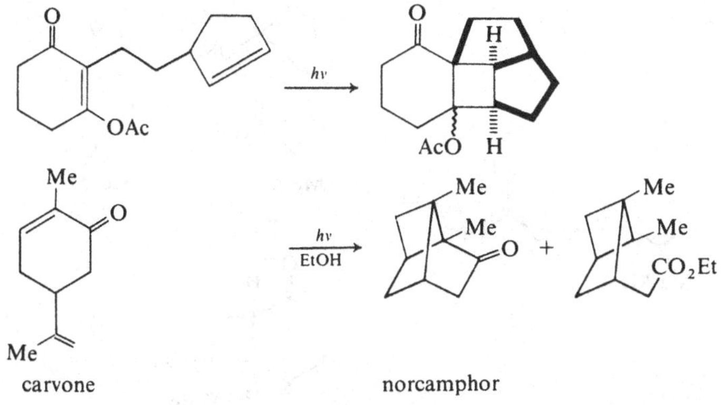

Scheme 4.8

yield. Irradiation with a krypton ion laser at the tail of the n–π* band results in rapid conversion of carvone into norcamphor in high yield (*c*. 90%) and this is one of an increasing number of results where specific irradiation enhances the cleanness of a reaction. The enyne cycloaddition shown below is the first intramolecular example to provide a fused cyclobutene derivative.

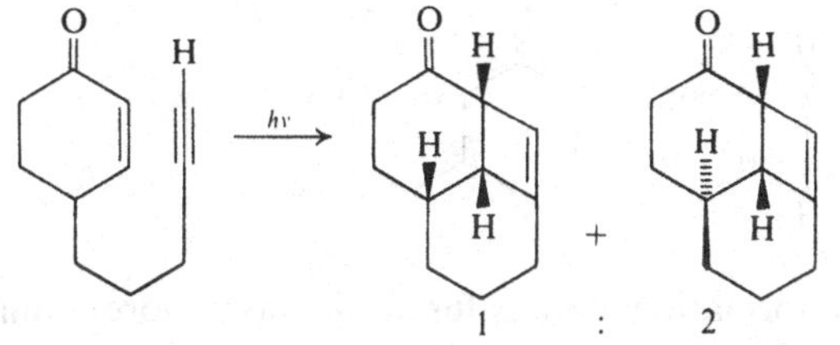

Occasionally *trans*-bicyclic products are formed as illustrated by reaction of the electron-rich dimethoxyethene with 4,4-dimethylcyclo-hexenone. The formation of a *trans*-adduct at the C_2–C_3 bond of the enone is limited to cyclohexenones and larger ring enones. Triplet formation of the *cis*-adduct from the reaction of the dimethoxyethene with the cyclohexenone is quenched faster than that of the *trans*-adduct and the oxetane is quenched with an efficiency comparable with that for the *trans*-adduct. It is thought therefore that the oxetane and the *trans*-adduct are formed from the same triplet state and that the *cis*-adduct is formed from a different triplet state. The preferential quenching of the triplet state that gives rise to the *cis*-adduct allows for the selective synthesis of the *trans*-adduct while the more thermodynamically stable *cis*-isomer can be produced by thermal equilibration of mixtures of the adducts.

Reaction of cyclohexenone with 2-methylpropene gives both of the possible *cis*-fused cyclobutane adducts, but the major product is the *trans*-

fused ketone (15). The formation of *trans*-fused adducts is more common for reactions of enones with electron-rich alkenes and is not observed for reactions of cyclopentenones. While it is possible that the *trans*-fused products arise by initial *cis–trans* isomerisation of the enone, it is more likely that an exciplex with a highly twisted character which may even approach a *trans*-structure in geometry is involved. The formation of the monocyclic alkene products in this latter reaction suggests the intermediacy of diradicals.

The reaction of enones with alkenes is known to be effected by n–π* excitation of the enone since the reaction can be effected under conditions where the alkene does not absorb light. The reaction is thought to involve exciplex formation and/or the intermediacy of a diradical and this is shown diagramatically in scheme 4.9. If the reaction proceeds directly to a diradical intermediate the diradical will be a triplet and the regioselectivity of reaction with an unsymmetrical alkene will be determined from the relative stability of the possible diradical intermediates. A radical centre adjacent to the carbonyl group will be less stable than one at C_3 and this

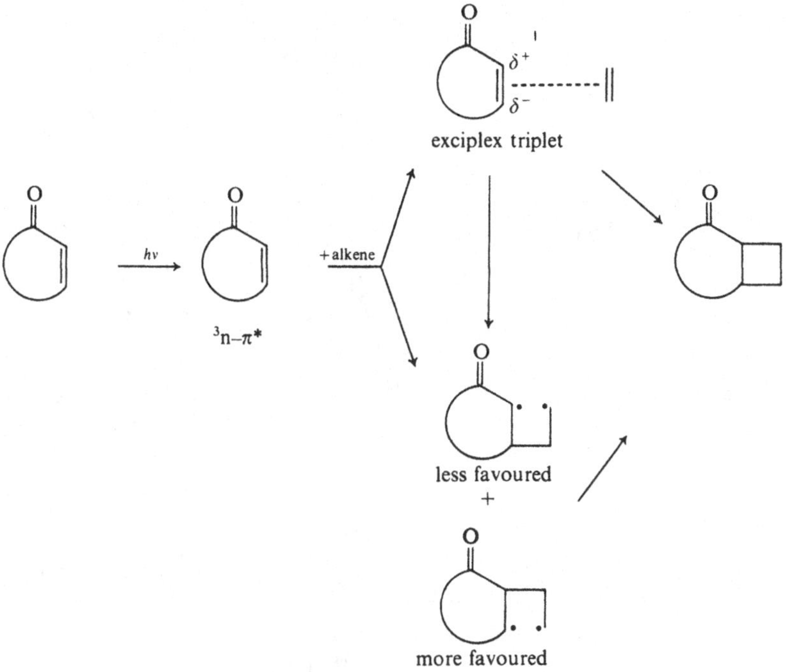

Scheme 4.9

dictates that bond formation should first occur at C_2 of the enone. If on the other hand the reaction proceeds from an exciplex, the exciplex will have a considerable amount of charge-transfer character. A triplet excited enone is considered to have the polarity depicted in the scheme 4.9. This will dictate the orientation of the alkene in the exciplex and therefore the regioselectivity of the adduct if formed directly from the exciplex. Alternatively the exciplex will partition to diradical intermediates and in this way dictate the regioselectivity of adduct formation. For the reactions of cyclopentenones where only *cis*-adducts are formed the reaction regioselectivity increases with increasing nucleophilic character of the alkene. Reaction of cyclopentenone with ethoxyethene affords (16) as the major product and this adduct can be envisaged as arising from attack of enone C_2 as electrophile on the unsymmetrical alkene. The favoured diradical or dissymmetric exciplex thereby determines the reaction product. The reaction with dimethoxyethene is even more regioselective and this is consistent with more directional character in the exciplex–diradical intermediate reflecting the effect of the extra ether substituent on the alkene. The stereoselectivity of the alkene in the addition is largely dictated by steric factors and the adduct (17) (scheme 4.10) is the dominant product.

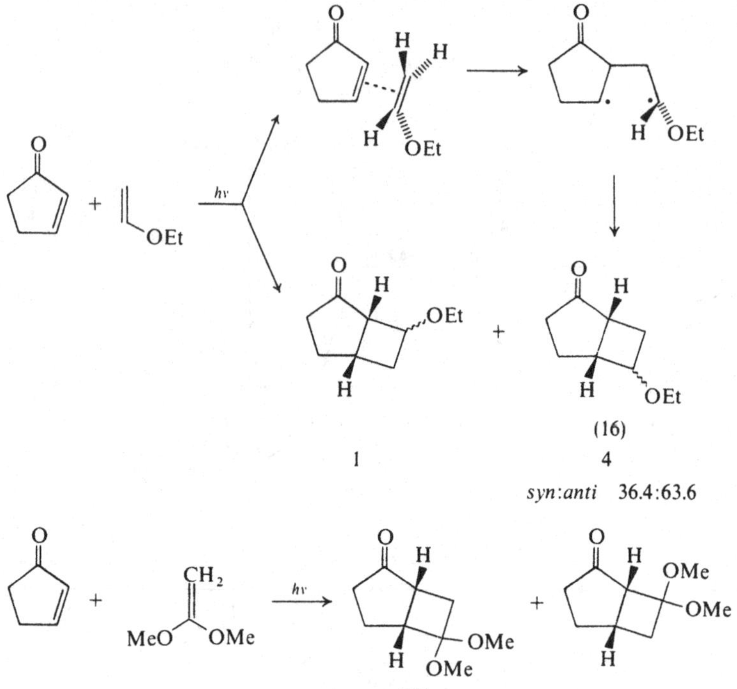

(16)

1   4

syn:anti 36.4:63.6

(17)

Scheme 4.10

In contrast to the reaction of cyclopentenone, photolysis of cyclohex-enone with dimethoxyethene gives largely the *trans*-bicyclic isomer. While electron-rich ethenes react more regioselectively and more rapidly than

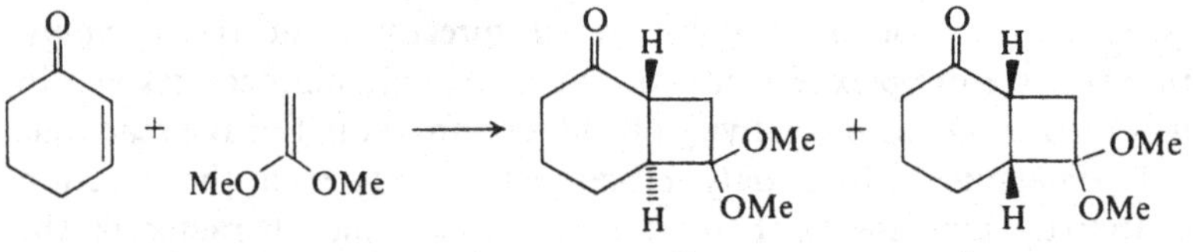

7   :   3

electron-deficient alkenes, cycloaddition occurs with loss of stereochemistry of the ethene fragment. Cyclohexenone reacts with both *cis*- and *trans*-butene to produce virtually identical mixtures of the three possible bicyclo[4,2,0]octanones demonstrating that the stereochemical integrity of the alkene is lost in the reaction. This suggests the intermediacy of diradicals, at least for electron-rich alkenes, which are of sufficient lifetime for bond rotation to occur and for loss of memory at the radical centre formed from the alkene. For attack at C_2 or C_3 of the enone, a pair of diastereoisomeric diradicals can be formed and the ratio of the diastereoisomeric radicals in each pair would be expected to vary depending upon whether the alkene was initially *cis* or *trans*. The similarity of the product mixtures cannot therefore be accounted for simply by assuming that the diradical intermediates have time for bond rotation to occur before ring closure.

Photochemical cycloaddition of *cis*- and *trans*-1,2-dichloroethene to cyclopentenone gives three of the four possible *cis*-fused adducts (scheme 4.11). The ratio of adducts is such that the reaction can be explained only if

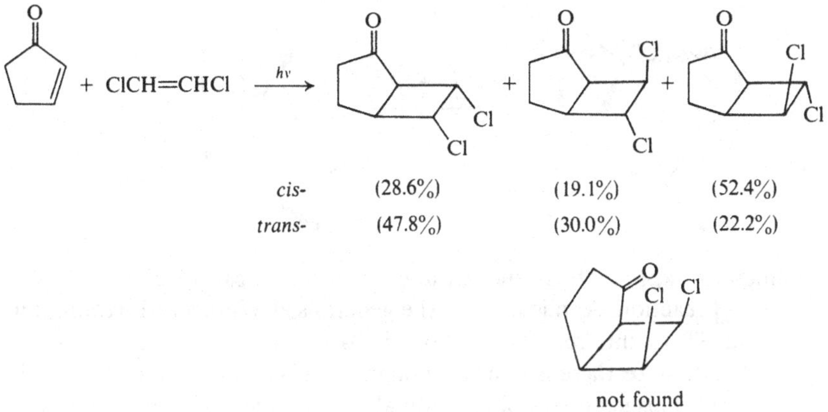

+ ClCH=CHCl
hν
cis- (28.6%) (19.1%) (52.4%)
trans- (47.8%) (30.0%) (22.2%)
not found

Scheme 4.11

bond formation occurs initially not to C_2 of the enone – which is the usual case – but to C_3. Several dienones and *para*-quinones undergo intermolecular $[_\pi 2+_\pi 2]$ dimerisations but, unlike the reactions of the αβ-unsaturated cyclic ketones discussed above, these reactions occur with predominant *cis*-cisoid-*cis* stereochemistry as shown for the formation of (18). In many instances the initial product undergoes an internal $[_\pi 2_s+_\pi 2_s]$

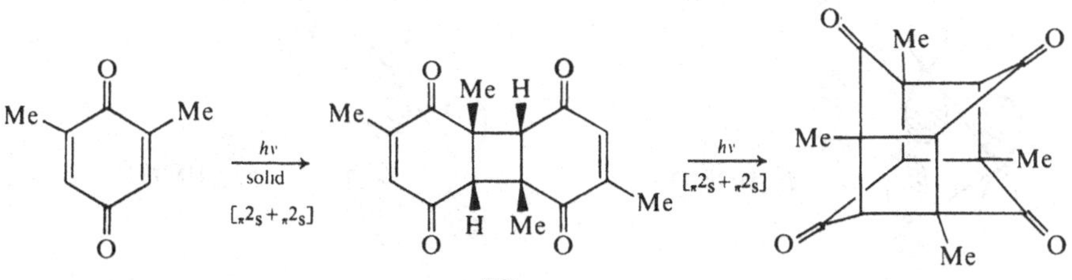

hν
solid
$[_\pi 2_s+_\pi 2_s]$
hν
$[_\pi 2_s+_\pi 2_s]$

(18)

cycloaddition to afford the cage dimer. Molecules of this type are of current interest as energy stores with metal catalysed opening of the cage to release the 'trapped sunlight'.

4.4   Other cycloadditions[1-5,7,8,14-20]

The Diels–Alder $[_\pi 4 + _\pi 2]$ cycloaddition is the best known cyclo-addition.[14-20] When the addends lie on top of each other the maximum

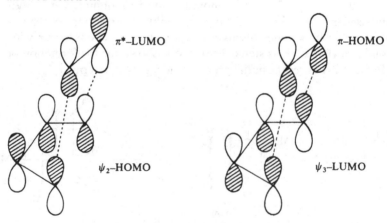

symmetry of approach is attained and the reaction can be classified as a $[_\pi 4_s + _\pi 2_s]$ reaction. Application of the generalised Woodward–Hoffmann rule[1] (p. 38) to this reaction dictates it to be symmetry allowed in the ground state since there is one component that is counted, namely the $_\pi 2_s$ component. This geometry of reaction allows for favourable mixing of the highest occupied and lowest unoccupied orbitals in going to the transition structure and thus meets the criterion of frontier orbital theory[4,5] for an allowed reaction.

π^*–LUMO

π–HOMO

ψ_2–HOMO

ψ_3–LUMO

$[_\pi 4_s + _\pi 2_s]$ cycloaddition

The transition structure for this reaction is of the Hückel type[15] since there are no sign inversions in the cyclic array (see section 2.2) which

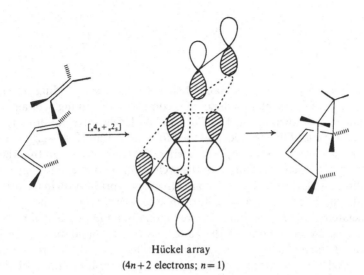

Hückel array
($4n + 2$ electrons; $n = 1$)

contains six electrons, *viz.*, $(4n + 2)$ electrons with $n = 1$. As such the transition structure has aromatic character and is formed in a ground state reaction (see table 2.3). The concept of aromatic transition structures was first proposed by Evans[16] in 1939 and its more recent revival has been due to Dewar.[15] As discussed in chapter 2 the transition structure of a reaction is drawn as a series of overlapping s- and p-orbitals with positive and negative phases inserted so as to minimise the number of phase changes in the participating orbitals. The number of sign inversions in the cyclic array and the number of electrons involved are counted. If the cycle which defines the stereochemistry of the reaction includes both lobes of a single p-orbital the sign inversion across the lobes is not counted. The cyclic array is defined as Möbius or Hückel depending on whether the number of countable phase inversions in the array is odd or even respectively. If the reaction is a Hückel system and also has $(4n + 2)$ delocalised electrons, the reaction has an aromatic transition structure and is thermally favoured. On the other hand, if the system is of the Möbius type and the number of electrons is $(4n)$ then the reaction is also said to have an aromatic transition structure and is thermally favoured. The converse is that a $(4n)$ Hückel and a Möbius $(4n + 2)$ system is thermally unfavourable. For photochemical reactions the rules are reversed (see table 2.3).

To construct a correlation diagram[1] for the allowed $[_\pi 4_s + _\pi 2_s]$ reaction the symmetry of reaction is defined as a plane bisecting the diene and the alkene as shown in fig. 4.8. The σ-, σ^*-, π- and π^*-orbitals most involved in the reaction are defined as symmetric or antisymmetric with respect to this plane and are arranged in order of approximately increasing energies. For the product cyclohexene, the delocalised orbital combinations of σ_1 and σ_2,

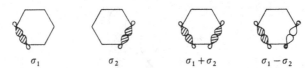

σ_1 σ_2 $\sigma_1 + \sigma_2$ $\sigma_1 - \sigma_2$

namely $\sigma_1 + \sigma_2$ and $\sigma_1 - \sigma_2$, and their respective antibonding levels are considered. The correlation diagram is completed by joining orbitals of like symmetry as shown in fig. 4.9. The bonding levels correlate directly with bonding levels of the product and therefore the thermal $[_\pi 4_s + _\pi 2_s]$ reaction has no orbital symmetry imposed barrier. This does not mean that there is no activation energy required for the cycloaddition process. Energy will be required to effect the changes accompanying rehybridisation in the orbitals, bond length changes, and distortion and change in the bond angles associated with the reaction. An exceptionally large number of reactions occurring by interaction of the addends in the ground state have been reported and it is generally accepted that the majority of these reactions are concerted.[17,18] The two possible $[_\pi 4_s + _\pi 2_s]$ reaction pathways shown in fig. 4.10 will be differentiated from a consideration of steric factors and secondary orbital interactions that occur in developing the transition structure and the preferred pathway will depend on the nature of the Y and X substituents.

This $[_\pi 4_s + _\pi 2_s]$ mode of addition, while by far the most common, is not the only relative geometry of addends from which reaction can occur and the alternatives are shown in fig. 4.10. The relative geometry of the appropriately substituted addends as they react dictates a particular stereochemistry in the reaction product. The principle of conservation of

Fig. 4.8. Maximum symmetry approach of ethene and butadiene.

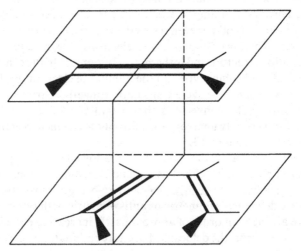

orbital symmetry in the course of the reaction determines which of the higher occupied molecular orbitals and lower unoccupied molecular orbitals transform to the product higher occupied and lower unoccupied molecular orbitals. The symmetry-allowed nature or otherwise of each of the reaction courses can be determined by using the Woodward–Hoffmann generalised rule (see p. 38).

For any particular addends the choice of reaction geometry will result from an interplay of factors.[2,7,8] Steric considerations are of particular importance, firstly in differentiating the sterically reasonable from the sterically impossible reaction geometries and secondly, where there is more than one sterically sensible geometry which represents an allowed process,

Fig. 4.9. Correlation diagram for $[_\pi 4_s +_\pi 2_s]$ cycloaddition.

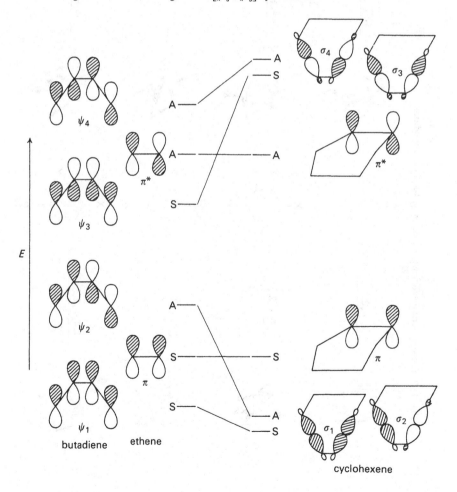

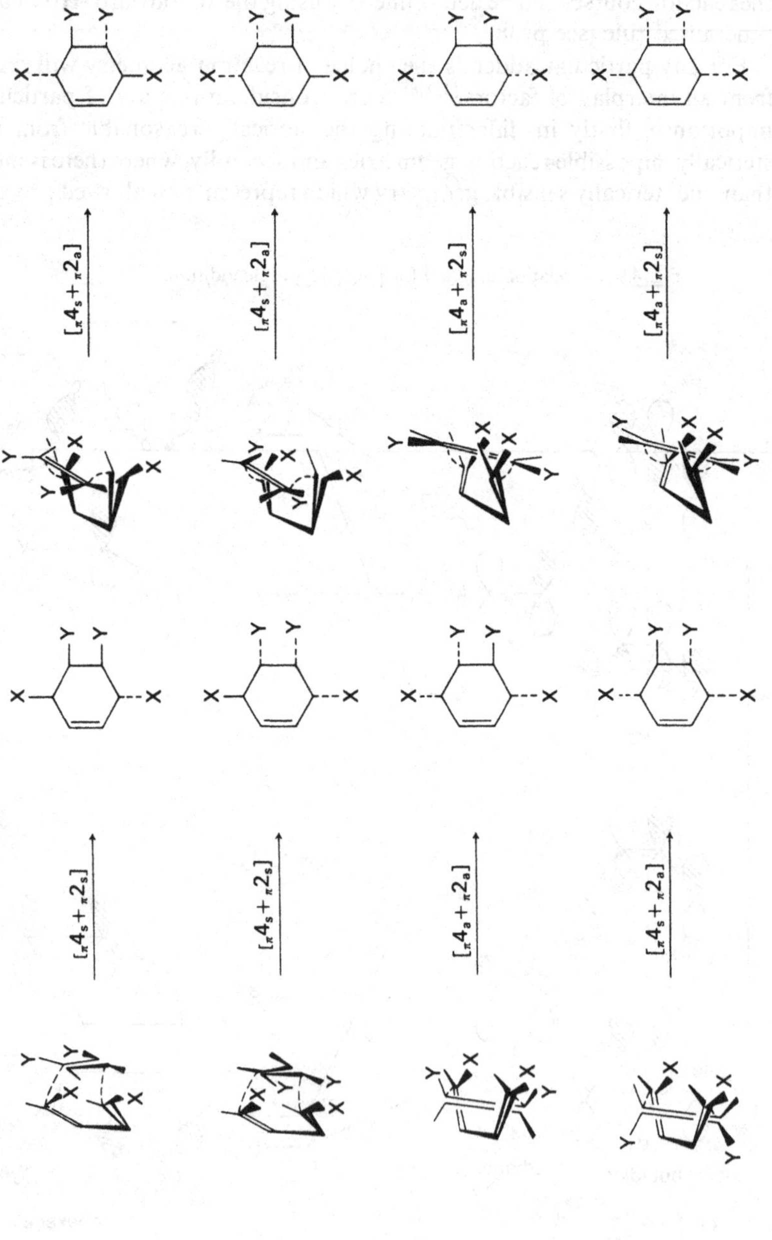

Fig. 4.10. The stereochemical consequences of $[_\pi 4 + _\pi 2]$ cycloaddition.

in determining the partition of the reaction between these pathways. For example, the $[_\pi 4_a +_\pi 2_s]$ reactions shown in fig. 4.10, even though orbital symmetry allowed from the ground state, would be considered unlikely owing to the unfavourable contortions of the reactants necessary for this reaction to occur. The $[_\pi 4_a +_\pi 2_s]$ and the $[_\pi 4_s +_\pi 2_a]$ reactions are not favoured thermally but are allowed photochemically. The first excited state for each of these reaction geometries correlates with the first excited state of the product and therefore it would be from a photochemical reaction that one would look for such a reaction geometry. Steric considerations and secondary orbital interactions[7] would be important in determining any partition of a reaction between these routes.

A state correlation diagram which collates not only the ground state reaction but also higher state reactions is shown in fig. 4.11. The diagram shows the symmetry-imposed barrier for the first excited state reaction; indeed simple $[_\pi 4_s +_\pi 2_s]$ Diels–Alder reactions are not observed photochemically. However, the $[_\pi 4_s +_\pi 2_a]$ (shown in fig. 4.10 and below) and $[_\pi 4_a +_\pi 2_s]$ modes of cycloaddition are allowed in the excited state and a

Fig. 4.11. Electron state diagram for formation of cyclohexene from butadiene and ethene.

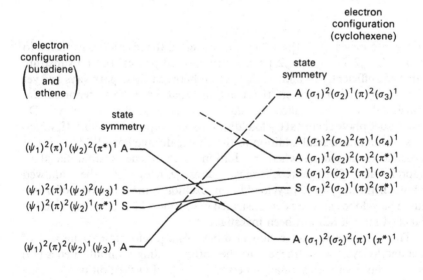

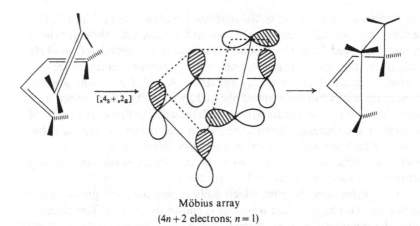

Möbius array
($4n+2$ electrons; $n=1$)

number of reactions that fit the criteria defined in this way have been observed particularily for intramolecular cases.

Intramolecular $[_\pi 4 +_\pi 2]$ cycloadditions are also typified by crossed cycloaddition ($x[_\pi 4 +_\pi 2]$) as illustrated by the conversion of *cis*-hexa-1,3,5-

$$\overset{CH_2}{\underset{CH_2}{}} \xrightarrow[\substack{[_\pi 4_s +_\pi 2_a] \\ or \\ [_\pi 4_a +_\pi 2_s]}]{hv} \quad = $$

triene into bicyclo[3,1,0]hex-2-ene. If concerted the reaction is expected to be a $[_\pi 4_s +_\pi 2_a]$ or $[_\pi 4_a +_\pi 2_s]$ symmetry-allowed process but no substrate labelled sufficiently to allow a distinction between these pathways has yet been examined. However, the reactions that have been reported are consistent with the allowed modes of cycloaddition. Vitamin D_2 undergoes photochemical cycloaddition to give suprasterol I and II, which arise from antarafacial addition to the C_7–C_8 double bond.[19,20] It is not possible to determine whether addition to the diene is antarafacial or suprafacial. The stereochemical conseqeunces of the allowed photochemical $[_\pi 4 +_\pi 2]$ cycloaddition, excluding the formation of *trans*-fused bicyclohexenes, are demonstrated in scheme 4.12 and as yet a suitably labelled system has not been investigated.

The spontaneous conversion of dibenzo[*a,d*]cyclooctatetraene and of octamethylcyclooctatetraene into the corresponding semibullvalenes (19) are examples of the thermally-allowed $[_\pi 4_a +_\pi 2_a]$ cycloaddition.

Cyloadditions are not limited to neutral molecules and a number of reactions are known where an addend is an ion. Such reactions are shown in scheme 4.13 and can be regarded at least formally as $[_\pi 4 +_\pi 2]$ reactions.

Cycloadditions which involve more than six electrons are uncommon.[3]

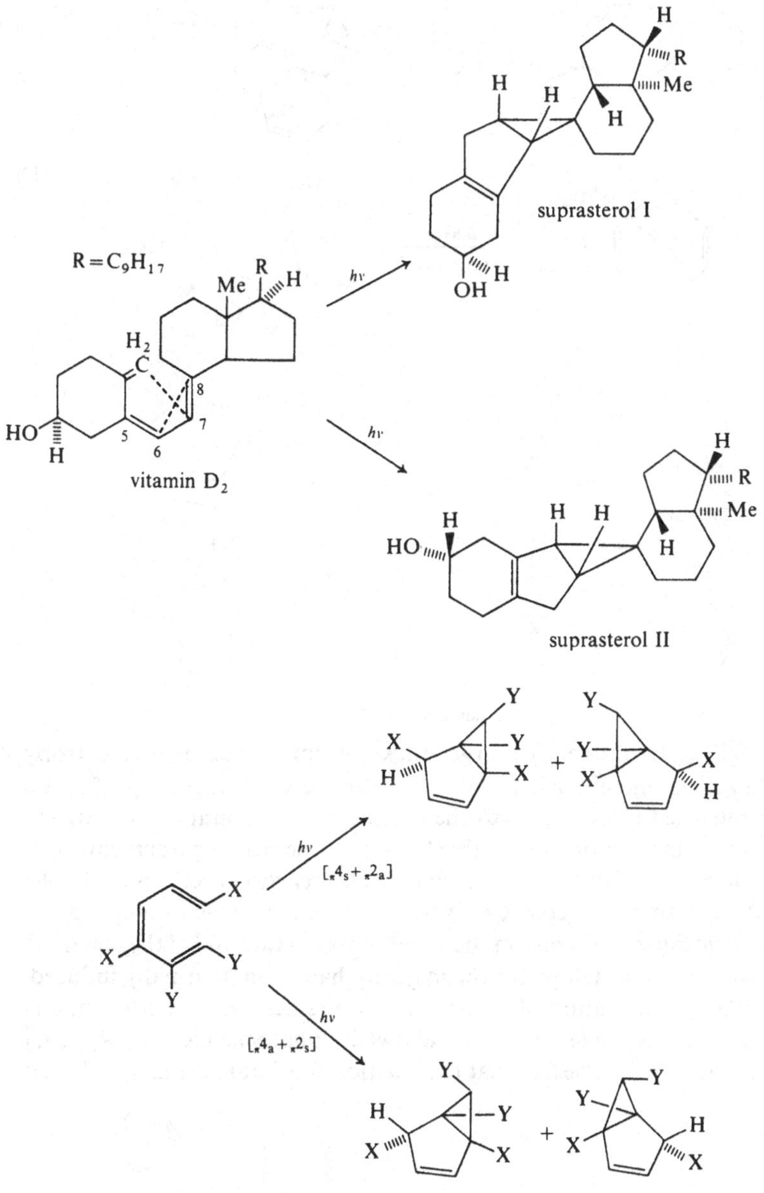

Scheme 4.12

Nevertheless, the success of the Woodward–Hoffmann rules[1] in defining reactions that are orbital symmetry allowed has prompted the search for reactions which involve more than six electrons. Concerted reactions of

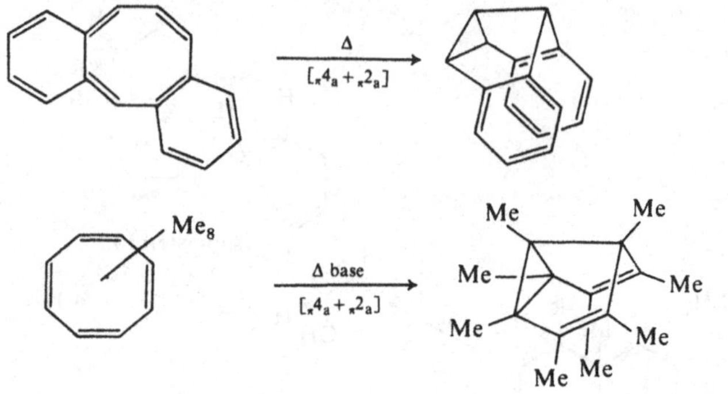

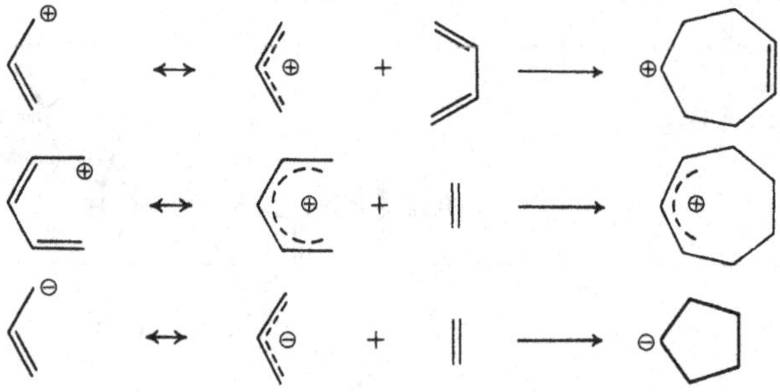

Scheme 4.13

open chain polyenes to give large rings are not favoured from entropy considerations and if such reactions are to be encouraged appropriate constraints need to be built into the molecules. For example, the entropy factor can be made more favourable by having the reacting fragments held in a single molecule whereupon intramolecular reaction becomes probable. Nevertheless, there are comparatively few reactions which involve more than six delocalised electrons in the transition structure and of the reactions that have been reported by far the majority have been thermally induced.

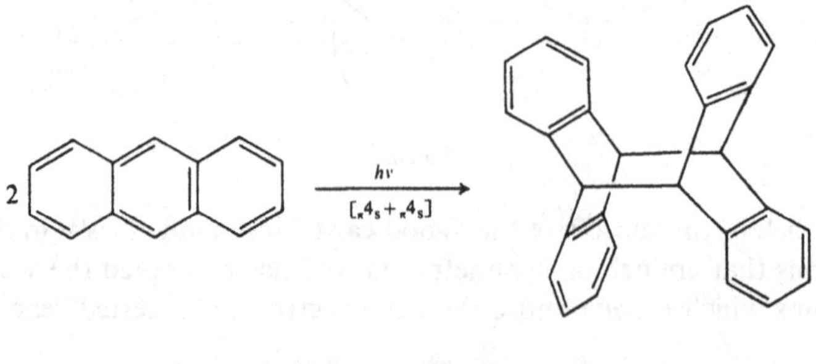

The photodimerisation of anthracene occurs across the 9,10-positions and is an example of an allowed photochemical $[_\pi4_s+_\pi4_s]$ cycloaddition.[19,20] The fact that the reaction is orbital symmetry allowed (see section 4.6) does not prove, in any way, that the reaction is concerted and a considerable amount of detailed mechanistic work is necessary before such a conclusion can be stated. 2-Pyridones are also known to undergo

$[_{\pi}4_s+_{\pi}4_s]$ dimerisation as does 4,6-diphenyl-α-pyrone. This latter compound undergoes photocyclisation in the solid state to give the $[_{\pi}4_s+_{\pi}4_s]$ *anti*-dimer as the sole product of reaction. In dilute solution only

$[_{\pi}2_s+_{\pi}2_s]$ cyclodimers are formed and this is another example of the interesting change in reaction course when the solid as opposed to a solution of the compound is subjected to irradiation. The addition of butadiene to benzene is at least formally another example of a $[_{\pi}4_s+_{\pi}4_s]$ reaction.

The $[_{\pi}6_s+_{\pi}2_s]$ cycloaddition, while allowed in the excited state, is rare; a product consistent with such a reaction is found in the photolysis of tropone, but since it is only one of a number of products it could arise equally well from a non-concerted reaction.

The reaction of tropone with 2-methylpropene is formally a $\left[_\pi 8 +_\pi 2\right]$ reaction and is thought to involve addition of the n–π* triplet excited state of tropone to the alkene. There are several examples of thermal reactions which involve ten electrons in the transition structure and these can generally be regarded as involving suprafacial attack on each of the reacting species.

Irradiation of a dilute acid solution of tropone gives a dimer resulting

from an allowed $\left[_\pi 6_s +_\pi 6_s\right]$ cycloaddition. However it is unlikely that this reaction is concerted. In like manner, 2-chlorotropone gives the $\left[_\pi 6_s +_\pi 6_s\right]$ dimer as well as the $\left[_\pi 4_s +_\pi 2_s\right]$ dimer.

4.5   Cycloadditions of benzene and its derivatives[3,10,21–23]

In contrast to the ground state chemistry of benzene, which has been explored in great detail over a long period, the study of the photochemistry of simple aromatic systems is more recent in origin.[3,21,22] The most noted reactions of benzene and its derivatives are isomerisation, which is discussed in section 2.2, dimerisation and the formation of addition products. The synthetic application of the rich photochemistry of aromatic systems is yet to be fully exploited.

The addition of an alkene, diene, alkyne, amine, alcohol or carboxylic acid to benzene[10] can occur across the *ortho-*, *meta-* and *para-*positions to give three distinct products. For alkene addends *ortho-*, *meta-* and *para-*cycloadditions are all stereospecific with respect to the alkene component as shown in scheme 4.14. The reactions generally occur by excitation of the

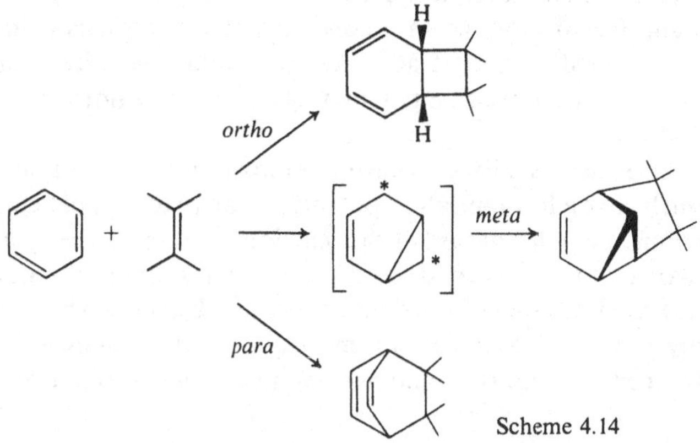

Scheme 4.14

benzene chromopore, but it is possible to have initial excitation of the alkene addend or, in some cases, excitation of a charge-transfer complex formed between addends. All three types of addition, *ortho*, *meta*, and *para*, can occur from the singlet excited state of the aromatic chromophore.

The π-orbitals of benzene ($\psi_1-\psi_6$, fig. 4.12) can interact with the π- and π^*-orbitals of ethene in different ways to give *ortho-*, *meta-* and *para-*adducts.[10,21,22] For *ortho-*addition a correlation diagram can be constructed as shown in fig. 4.13. The molecular orbitals of benzene are placed on one side of the diagram and the σ- and π-orbitals of the product bicyclooctadiene on the other and in order of increasing energy (cf. section 2.2). The orbitals are classified with respect to a symmetry plane which bisects the addends (and is equivalent to a vertical plane through the benzene orbitals of fig. 4.12) and the correlation diagram completed in the usual way. The S_1 excited state of benzene (the $^1B_{2u}$ state) arises from the forbidden transition at 254 nm ($\varepsilon = 20.4 \, \text{m}^2 \, \text{mol}^{-1}$) and involves the promotion of an electron from ψ_2 (or ψ_3) to ψ_4 (or ψ_5). Since $\psi_2\psi_4$ and $\psi_3\psi_5$ are equally a part of the wave function of S_1 benzene, *either* can serve to establish a favoured correlation with the reaction product. Although more complex than the earlier (chapter 2) correlations it can be seen from fig. 4.13 that the photochemical *ortho-*addition of ethene to benzene is unfavourable. Analogous correlation diagrams can be constructed for *meta-*addition (fig. 4.14) and for *para-*addition (fig. 4.15). Only *meta-*addition is an allowed process and this can occur in two distinct ways depending upon the timing of the various bond formation processes (fig.

4.14). *meta*-Addition involves either direct concerted cycloaddition or initial *meta*-bonding in the singlet excited state to give a 'prefulvene-type' intermediate (fig. 4.14) which may have polar character. The formation of bond (i) can be before bonds (ii) and (iii), at the same time as bonds (ii) and (iii), or subsequently to (ii) and (iii). The reaction is stereospecific and insensitive to proton donors and solvent. These factors indicate that at least bonds (ii) and (iii) are formed in a concerted step. Products from *meta*-addition are formed in almost all reactions – and often predominate – particularly when the ionisation potential of the alkene is not very different from that of benzene.

Cycloaddition across adjacent centres (*ortho*-addition) occurs largely for alkenes which have a low ionisation potential relative to that of benzene; in these reactions the polar nature of the reaction overcomes the symmetry-imposed barrier to a concerted process. Further support for the polar nature of the reaction comes from donor substituted ethenes where reaction is markedly promoted by polar solvent. *meta*-Addition does not exhibit such a solvent effect. The *ortho*- and *para*-photocycloadditions are formally

Fig. 4.12. Molecular orbitals of benzene.

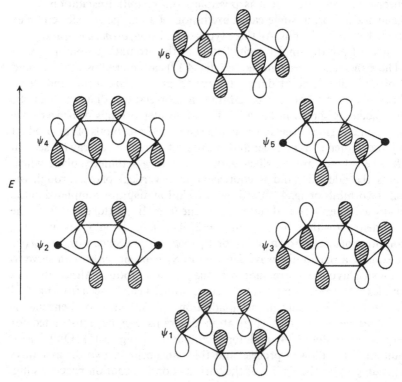

forbidden as concerted processes from interaction of S_1 benzene with S_0 alkene unless mixing with charge-transfer states occurs; intermediates with a high degree of ionic character are envisaged. This is supported by the effect of polar solvents and added proton donors on the course of reaction. When alkenes containing an isopropyldiene ($Me_2C{=}C$) moiety are employed, a non-cyclic 'ene-type' of reaction occurs (see section 5.1) and *para*-substitution is effected. The reaction is a nonconcerted process and is

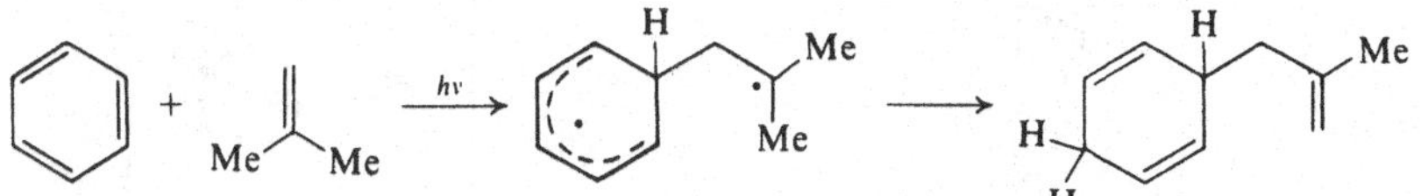

(20)

considered to proceed either via a diradical such as (20) or from an equivalent polarised exciplex. The reaction involves charge transfer by donation of an electron from the alkene to the excited benzene. It is not observed for anisole where the aromatic ring is electron rich.

Many reactions of benzene are thought to proceed via exciplexes. The formation of exciplexes and charge-transfer complexes may predispose a

Fig. 4.13. Correlation diagram for *ortho*-addition of ethene to benzene.

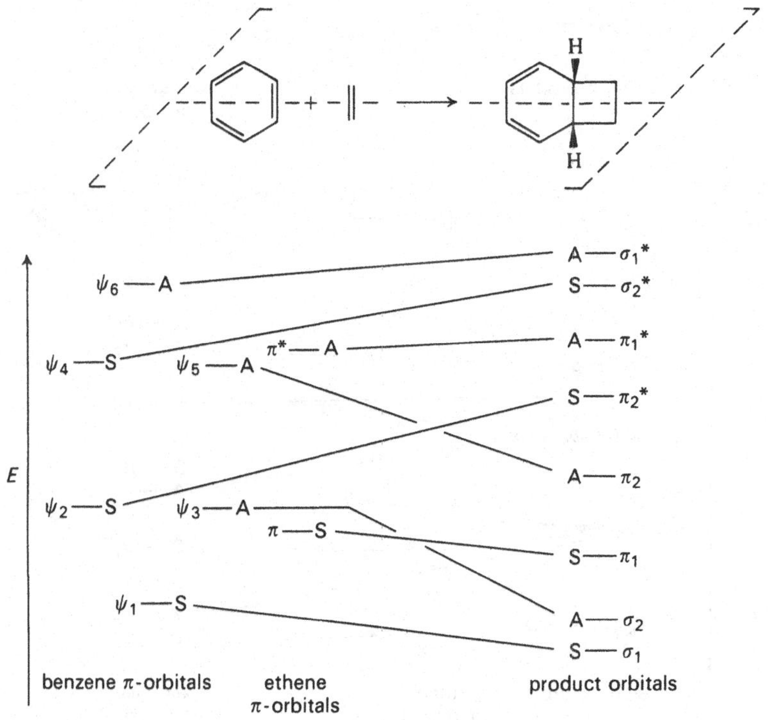

Fig. 4.14. Correlation diagrams for *meta*-addition of ethene to benzene (*a*) by initial formation of bond (i) to give a prefulvene diradical and (*b*) by initial formation of bonds (ii) and (iii).

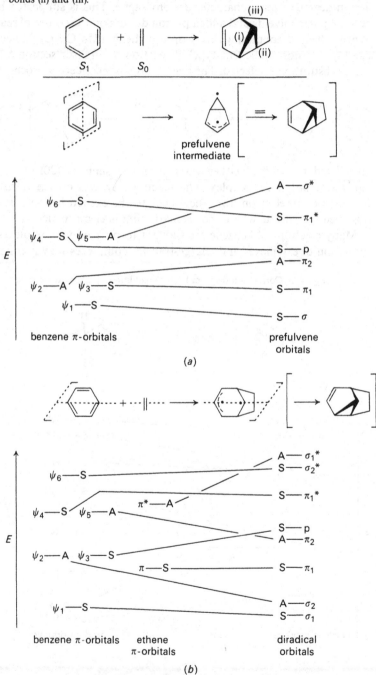

(*a*)

benzene π-orbitals

prefulvene
orbitals

(*b*)

benzene π-orbitals   ethene
π-orbitals

diradical
orbitals

pair of molecules to a particular relative orientation and consequently dictate subsequent reaction stereochemistry and reaction course. There is often little direct experimental evidence for exciplexes since reaction and dissociation usually compete with exciplex emission. However, emission has been observed from a number of exciplexes formed between benzene and alkenes and this number is growing rapidly. The formation of singlet exciplexes favours concerted and stereoselective cycloaddition processes. Moreover, since *ortho*-additions are forbidden, these reactions are most likely to occur in systems having a marked donor–acceptor character, and in systems where this is not the case *meta*-addition will dominate.

The primary product of a cycloaddition is often unstable to the conditions employed. For example, the photochemical addition of ethyne to benzene gives an *ortho*-adduct which undergoes facile thermal

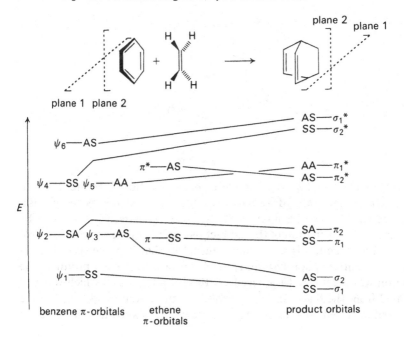

disrotatory electrocyclic ring opening of the cyclohexadiene unit to give cyclooctatetraene; alkynes do not undergo *meta*-addition to benzene. The

Fig. 4.15. Correlation diagram for *para*-addition of ethene to benzene.

reaction of benzene with 1,1-dimethoxyethene gives an *ortho*-cycloadduct which undergoes analogous electrocyclic ring opening and thus provides a useful synthetic approach to cyclooctatrienone (21). When the ethyne is

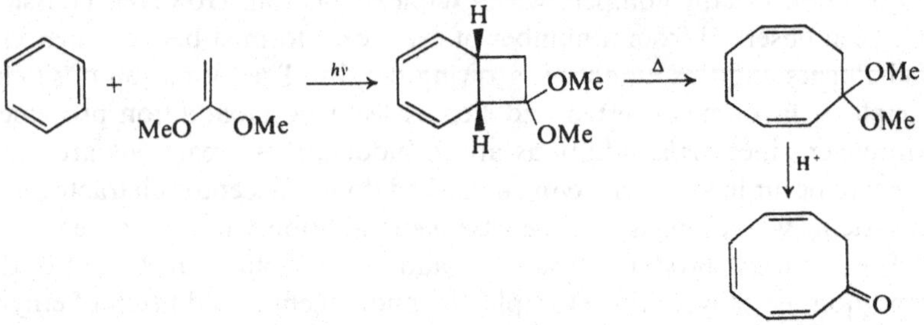

(21)

dimethyl ethynedicarboxylate, the initial adduct can be trapped before ring opening if tetracyanoethene is added to the reaction mixture (scheme 4.15).

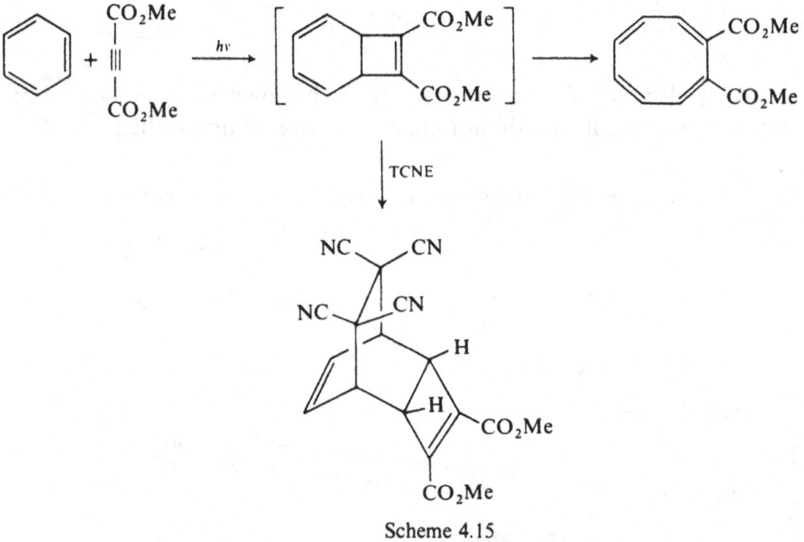

Scheme 4.15

Perhaps the best known example of *ortho*-addition is the reaction of benzene with maleic anhydride which was first reported in 1959 and yields a 1:2 adduct (23) as the final product of reaction (scheme 4.16). The mechanism for formation of this adduct has been a matter of considerable debate. This reaction is somewhat unusual in that it proceeds either on direct or sensitised photolysis and does not provide any *meta*-product. The reaction was presumed to proceed via the 1:1 adduct (22) which was not isolated from the reaction. Unsuccessful attempts were made to trap the expected intermediate (22) with tetracyanoethene (a known dienophile) by

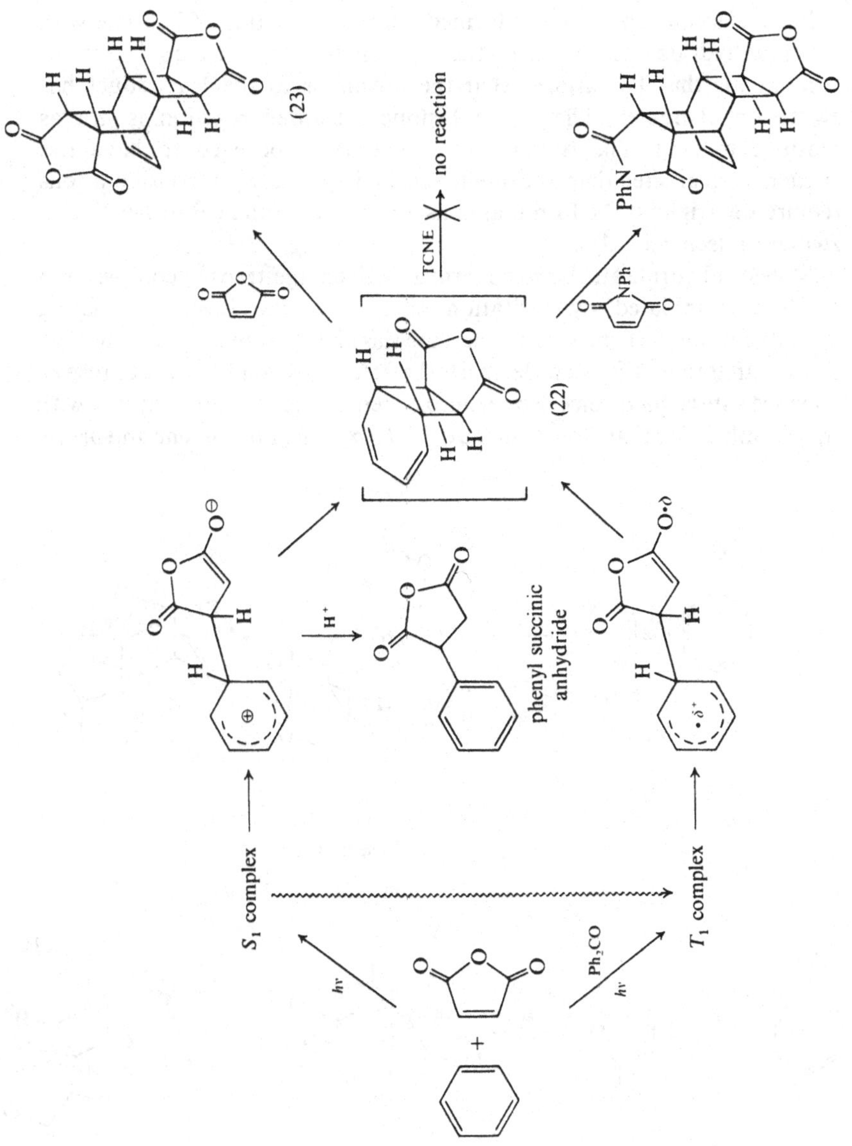

Scheme 4.16

Diels–Alder addition to the cisoid diene moiety. However, the initial adduct (22) has been diverted to (24) (scheme 4.16) with N-phenylmaleimide and in these circumstances the 2:1 adduct (23) with maleic anhydride is suppressed.

Phenylsuccinic anhydride is formed when the reaction of benzene with maleic anhydride is effected from the singlet manifold in acidic solvents. To account for this it is argued that the precursor to the 1:1 adduct has zwitterion character. The benzophenone sensitised reaction is not as sensitive to acid. The triplet excited species is believed to have less zwitterion character than the singlet excited species. Spin considerations require the triplet to be formulated as a polarised diradical rather than a zwitterion (scheme 4.16).

Maleic anhydride and benzene form a weak charge-transfer complex and reaction is initiated by excitation of this species. Electron-donating substituents on the benzene ring increase this charge-transfer character but, at the same time, they can decrease the rate of cycloaddition because of enhanced steric hindrance. For example, benzene reacts more rapidly with maleic anhydride than does toluene or *ortho*-xylene. For toluene and *ortho*-

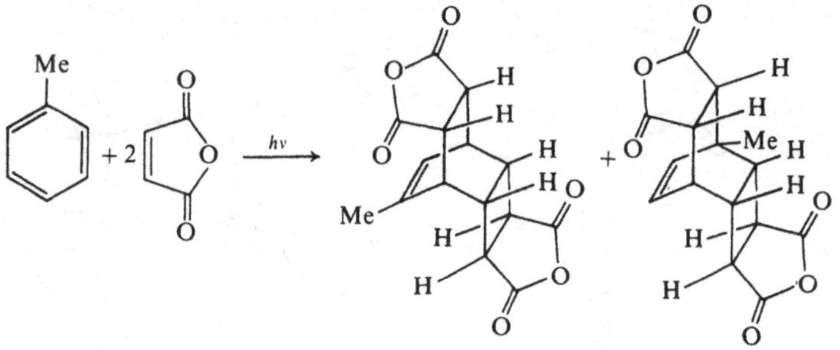

Sens 20°C   1:1
No sens 20°C   4:1

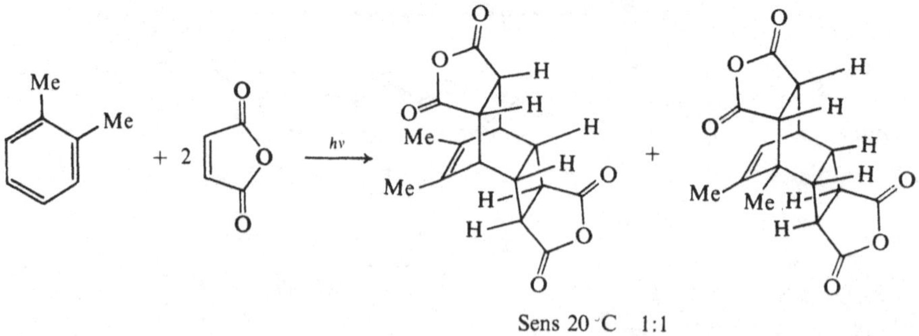

Sens 20°C   1:1

Scheme 4.17

xylene (scheme 4.17) the additions occur at positions remote from the alkyl substituents with the stereoselectivity being determined by the most favourable compromise between accessibility and donor character in the π-complex. This allows for maximum interpenetrability of the donor benzene and the acceptor maleic anhydride π-orbitals in the ground and excited states.

The analogous reaction of N-alkylmaleimides with benzene results in similar 2:1 adducts and for these reactions the initial photoproduct has been trapped with tetracyanoethene. Maleimide resembles maleic anhydride in its ability to complex with benzene but, unlike the reaction with maleic anhydride, the reaction is initiated by n–π* excitation of the maleimide rather than by excitation of the charge transfer complex.[23] The reaction of maleimide with anisole gives all three of the possible *ortho*-adducts as evidenced by isolation of the three 2:1 adducts.

For *ortho*-additions to benzenes there appears to be no general correlation between the orientation of the photoaddition and the orientation in the weak ground state complex. Consequently, the orientation in the excited state complexes appears to control the stereochemistry of addition. In the absence of overriding steric effects electron-deficient alkenes give mixtures of *exo*- and *endo*-adducts in *exo-*

apparently non-concerted processes. However, *ortho*-additions of electron-rich alkenes, such as enol ethers, give *exo*-adducts stereospecifically.

meta-Cycloadditions to benzenes have several orientational and stereo-chemical features that must be considered. In general the reaction proceeds via the prefulvene-type of intermediate. Substituents on the benzenoid ring will influence the site of 1,3-bridging in the aromatic ring, the regioselectivity of reaction, and the stereochemical outcome of addition to give *exo*- and *endo*-products.

The reaction of substituted benzenes (toluene, anisole, benzonitrile) with electron-rich alkenes such as ethyl vinyl ether give a mixture of *ortho*- and *meta*-products. The orientation of *meta*-addition with respect to the benzene ring is exclusively 2,6- for toluene and anisole. If this latter reaction were to proceed via a prefulvene intermediate the regiospecific formation of the 1-methoxy adduct (26; R = OMe) would be difficult to rationalise since

R = Me or OMe

(25)

(26)

formation of the necessary prefulvene intermediate requires 1,3-bond formation and places the methoxy substituent adjacent to a radical centre as shown for (25). It is likely, therefore, that the regiospecificity of the reaction is governed by the orientation of the addends in an exciplex.

The 1,3-photocycloadditions of a variety of cyclic and bicyclic alkenes to anisole, and of cyclobutene and cyclopentene to toluene, have each been reported to yield only one isomer which in many cases is the *endo*-isomer. This is shown for the reaction between cyclopentene and anisole where addition of the cycloalkene addend is to the 2,6-positions of anisole to give the *endo*-product (scheme 4.18).

Similar *meta*-addition reactions can be effected by photolysis in acetic

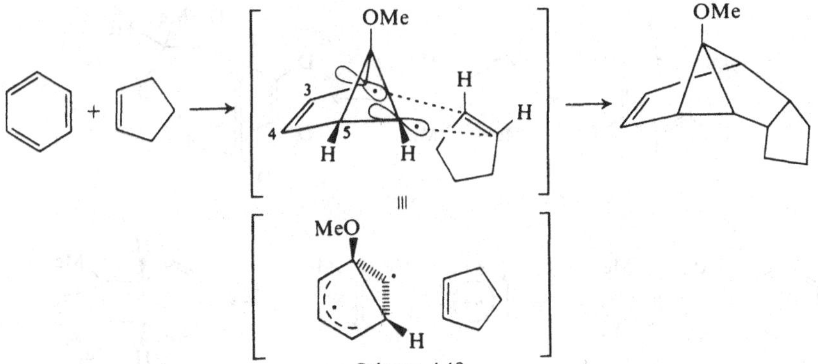

Scheme 4.18

acid by way of a protonated prefulvene intermediate. Support for such a species comes from attack by the acetate nucleophile from the less hindered face to deliver the bicyclo[3,1,0]hexenyl acetate (27). The intermediate can be formed by protonation of prefulvene or by direct phototransformation of protonated benzene.

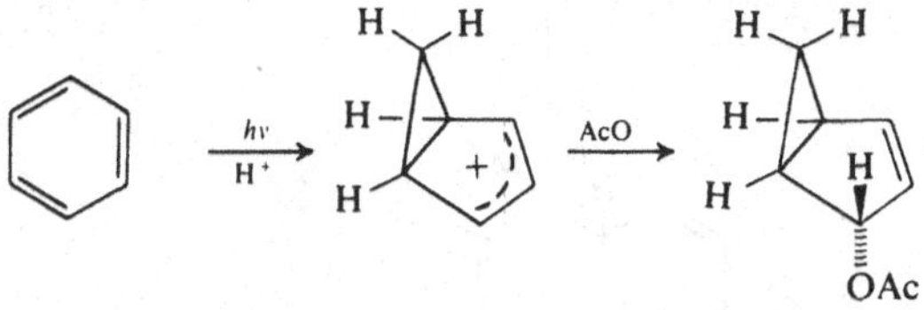

(27)

Addition across the 1,4-positions of benzene, *para*-addition, is much less common than *ortho*-addition. However, the two reaction types are believed to proceed by similar reaction mechanisms. The yield of *para*-products is often very low except when allenes are employed. Even so the *meta*-addition

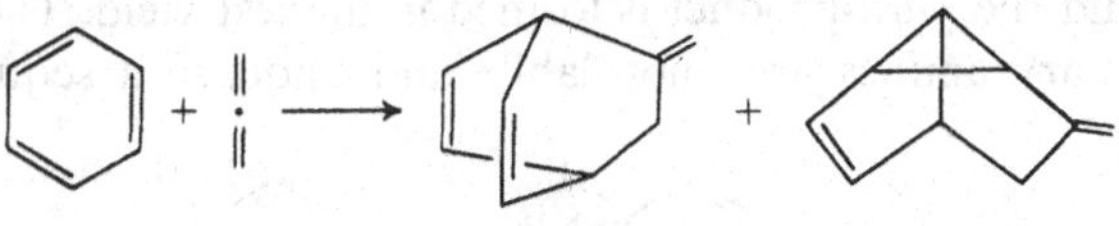

para-adduct   *meta*-adduct

product is also formed but with a lower quantum efficiency. A 1,4-ene type of reaction can also compete by a non-stereospecific process. This is illustrated for the reaction of deuterated benzene with dimethylbut-2-ene which gives a mixture of *cis*- and *trans*-cyclohexa-1,4-diene products. The hydrogen atom at C_1 is transferred as a proton. This follows from the appearance of deuterium at C_6 in the product when tetramethylethene is allowed to react with benzene in the presence of d^1-methanol. The normal *ortho*-, *meta*-, and *para*-cycloadducts are also formed in this latter reaction

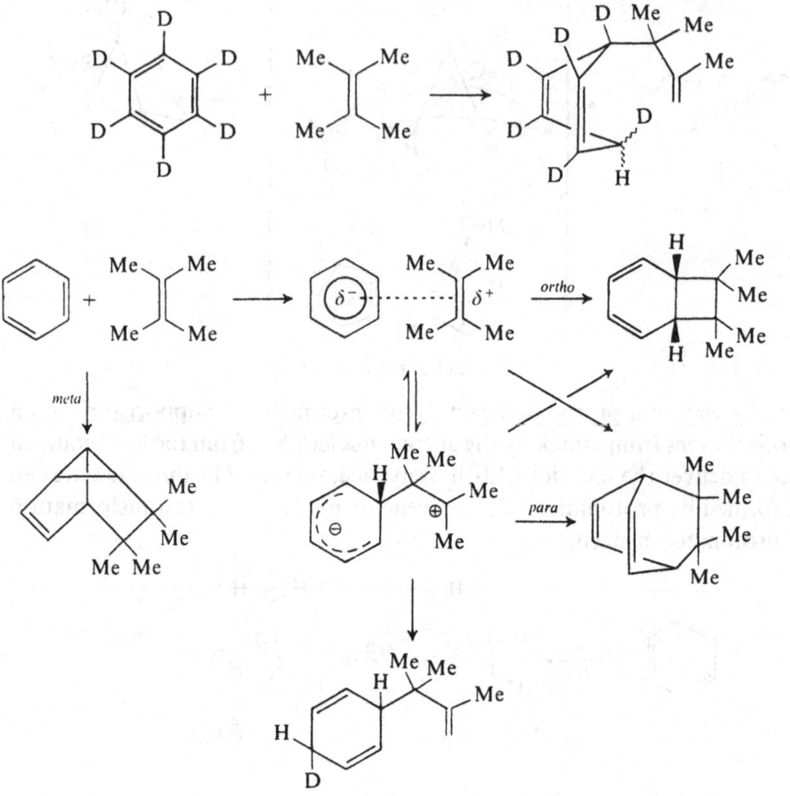

Scheme 4.19

and scheme 4.19 illustrates the possible reaction paths for their formation.

Primary and secondary aliphatic amines yield *ortho*- and *para*-adducts with benzene and the *para*-product is formed in highest yield. The *ortho*-adducts of primary amines are photolabile and undergo a sequence of

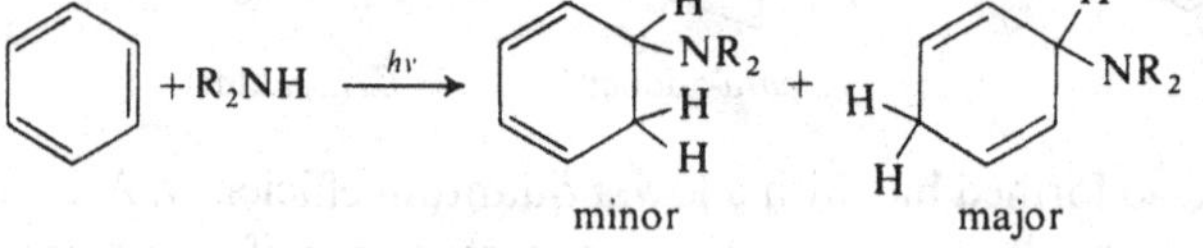

reactions which transform the compound into a cross adduct as shown in scheme 4.20. Tertiary amines are less reactive and require a proton donor such as methanol in order to add to the benzene ring; primary and secondary amines act as their own proton donors. The mechanism for the addition of triethylamine to benzene is shown in scheme 4.21 which, while complex, takes into account the known experimental data for this reaction.

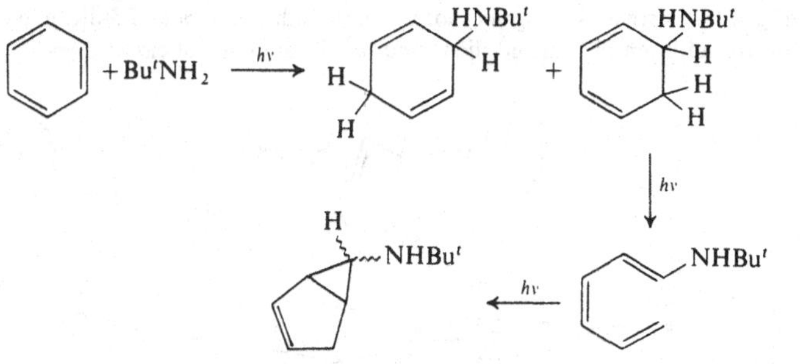

Scheme 4.20

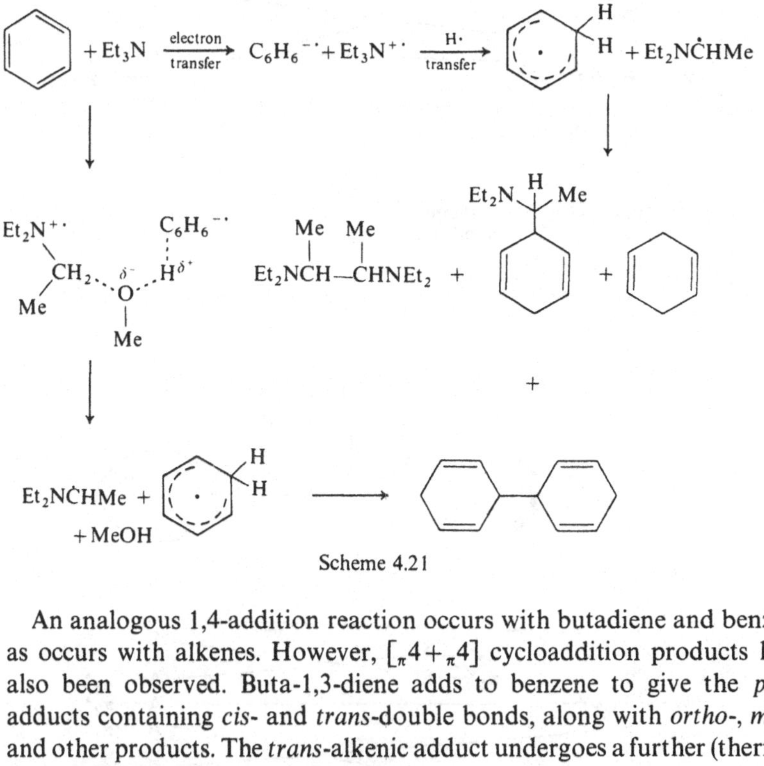

Scheme 4.21

An analogous 1,4-addition reaction occurs with butadiene and benzene as occurs with alkenes. However, $[_\pi 4 + _\pi 4]$ cycloaddition products have also been observed. Buta-1,3-diene adds to benzene to give the *para*-adducts containing *cis*- and *trans*-double bonds, along with *ortho*-, *meta*- and other products. The *trans*-alkenic adduct undergoes a further (thermal) addition reaction with butadiene to give (28) and a dimeric species (29) is observed which is thought to be formed by a thermally-allowed $[_\pi 2_s + _\pi 2_a]$ cycloaddition of the two initial products. The orbital symmetry correlation diagrams for the initial $[_\pi 4 + _\pi 4]$ reactions are shown in fig. 4.16. These diagrams indicate that the product from addition of transoid butadiene should be favoured since this conformer should add more readily than its cisoid counterpart. Whilst a $[_\pi 4 + _\pi 4]$ cycloaddition is an allowed

Fig. 4.16. Correlation diagrams for the *para*-addition of buta-1,3-diene to benzene (*a*) the addition of transoid diene and (*b*) the addition of cisoid diene.

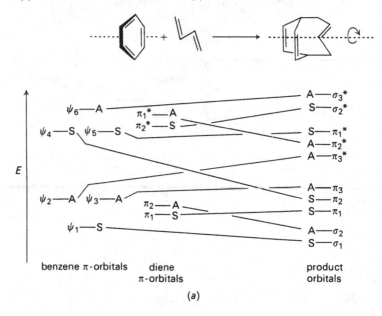

(*a*)

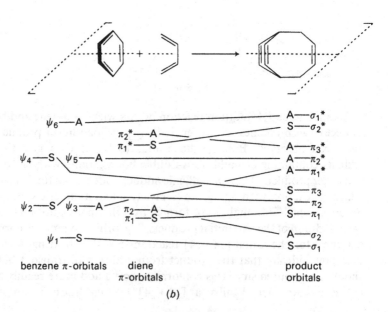

(*b*)

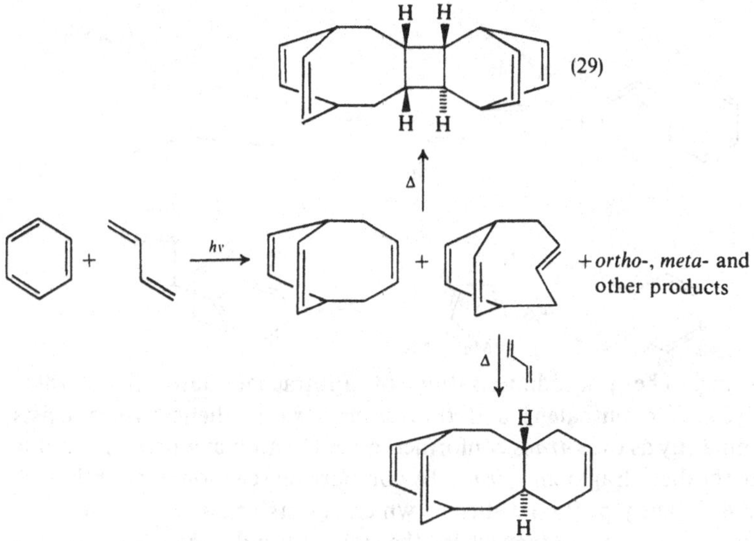
H H
H H
(29)
Δ
hν
+ ortho-, meta- and
other products
Δ
H
H

(28)

photochemical process there will be an unfavourable entropy of activation. This is particularly true for reactions involving alicyclic dienes which will make such a process less competitive than a multistep pathway in which the formation of one bond occurs in the primary photochemical process. The formation of a charge transfer complex between two ground state molecules, or an exciplex between the excited state of one molecule and the ground state of the other, can lower the barrier to a concerted addition. If the exciplex has a geometry such that the 1,4–1,4 positions of two species are proximate then such a reaction would be expected. The $[_\pi 4+_\pi 4]$ cycloaddition reaction can occur between two aromatic substrates. While such dimerisation has not been observed for benzene, it is well known amongst the acenes and is illustrated for 2-substituted naphthalenes and

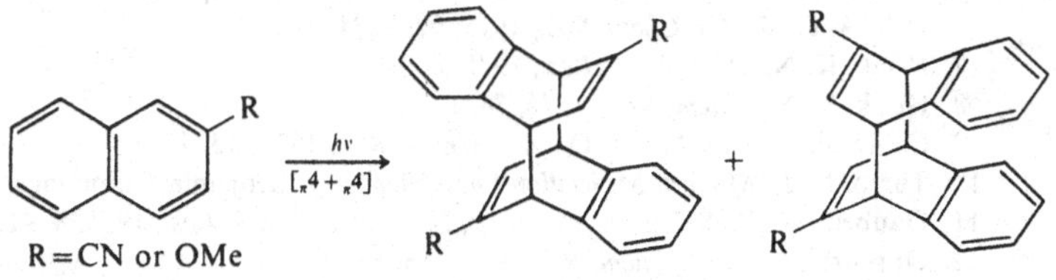
R
hν
$[_\pi 4+_\pi 4]$
R=CN or OMe
R
R
+
R
R

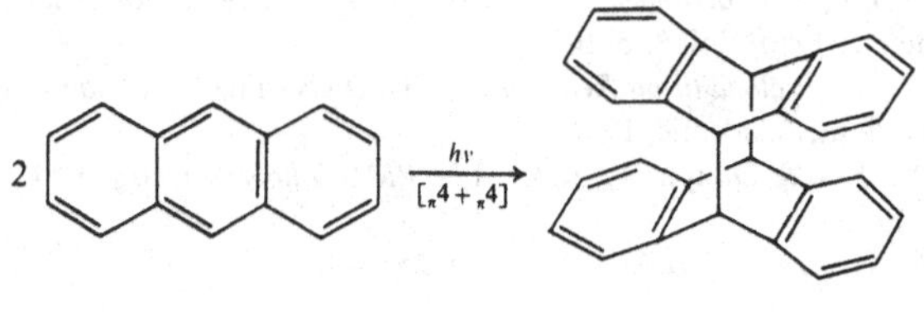
2
hν
$[_\pi 4+_\pi 4]$

anthracene. The photodimerisation of anthracene dates from 1866. Photolysis of naphthalene and *trans,trans*-hexa-2,5-diene (which exists predominantly as the s-*trans* conformer) gives the adduct where the double bond of the diene fragment is *trans*. By comparison reaction of naphthalene with 2,4-dimethylpenta-1,3-diene (which exists mainly as the s-*cis* conformer) gives the adduct where the double bond generated from the diene fragment is *cis*. Spectroscopic and kinetic data from these reactions implicate exciplexes in the quenching of the singlet excited state of the naphthalene. The results are consistent with the concerted $[_\pi 4 +_\pi 4]$ addition of diene to aromatic.

References

1. Woodward, R. B. & Hoffmann, R., *The Conservation of Orbital Symmetry*, Verlag Chemie, Academic, 1970.
2. Gilchrist, T. L. & Storr, R., *Organic Reactions and Orbital Symmetry*, 2nd edn, Cambridge University Press, 1979.
3. Bryce-Smith, D., ed., *Photochemistry – a Specialist Periodical Report*, The Chemical Society, Vols. 1–14.
4. Fleming, I., *Frontier Orbitals in Organic Chemical Reactions*, Wiley, 1976.
5. Fukui, K., *Theory of Orientation and Stereoselection*, Springer-Verlag, 1975.
6. Day, A. C., *J. Am. Chem. Soc.*, 1975, **97**, 2431.
7. Houk, K. N., *Acc. Chem. Res.*, 1975, **8**, 361.
8. Houk, K. N., *Chem. Rev.*, 1976, **76**, 1.
9. Caldwell, R. A. & Creed, D., *Acc. Chem. Res.*, 1980, **13**, 45.
10. Turro, N. J., *Modern Molecular Photochemistry*, Benjamin/Cummings, 1978.
11. Dauben, W. G., Salem, L. & Turro, N. J., *Acc. Chem. Res.*, 1975, **8**, 41.
12. Oppolzer, W., *Acc. Chem. Res.*, 1982, **15**, 135.
13. Schuster, D. I., in *Rearrangements in Ground and Excited States*, ed. de Mayo, P., Academic, 1980, Vol. 3, p. 167.
14. Hamer, J., *1,4-Cycloaddition Reactions – The Diels–Alder Reaction in Heterocyclic Synthesis*, Academic, 1967.
15. Dewar, M. J. S. & Dougherty, R. C., *The PMO Theory of Organic Chemistry*, Plenum, 1975.
16. Evans, M. G., *Trans. Faraday Soc.*, 1939, **35**, 824.

17. Dewar, M. J. S. & Pierini, A. B., *J. Am. Chem. Soc.*, 1984, **106**, 203.
18. Dewar, M. J. S., *J. Am. Chem. Soc.*, 1984, **106**, 209.
19. Dauben, W. G. & Phillips, R. B., *J. Am. Chem. Soc.*, 1982, **104**, 355; **104**, 5781.
20. Dauben, W. G., McInnis, E. L. & Michno, D. M., in *Rearrangements in Ground and Excited States*, ed. de Mayo, P., Academic, 1980, Vol. 3, p. 91.
21. Bryce-Smith, D. & Gilbert, A., in *Rearrangements in Ground and Excited States*, ed. de Mayo, P., Academic, 1980, Vol. 3, p. 349.
22. Bryce-Smith, D. & Gilbert, A., *Tertrahedron*, 1977, **33**, 2459; 1976, **32**, 1309.
23. Bryce-Smith, D., Gilbert, A. & Halton, B., *J.C.S. Perkin Trans. I*, 1978, 1172.

5 Oxidation, reduction, substitution and elimination reactions

5.1 Incorporation of molecular oxygen[1-11]

Photochemical reactions are often drastically altered by the presence of molecular oxygen. Under photochemical conditions, oxygen can insert itself into a substrate with the formation of hydroxyhydroperoxides (which readily decompose to carbonyl compounds), peroxides, hydroperoxides and dioxetanes.

The course taken by a photooxygenation reaction frequently depends on the electronic state of the oxygen molecule involved. In the ground state, molecular oxygen is a triplet. The highest occupied molecular orbitals (HOMOs) are the degenerate π^*-orbitals each containing a single electron. This is termed the $^3\Sigma$ state. Rearrangement of the electron spins within the degenerate orbitals results in two singlet excited states. Pairing of the electrons in one molecular orbital results in the $^1\Delta$ state, and spin pairing in the different orbitals results in the $^1\Sigma$ state (fig. 5.1). The $^3\Sigma$ and $^1\Sigma$ states have the same symmetry. The electron distribution in these species is identical, but differs from that in the $^1\Delta$ state. The energy differences between $^3\Sigma-^1\Delta$, and $^3\Sigma-^1\Sigma$ (92.4 and 159.6 kJ mol^{-1} respectively) are considerably lower than those between singlet ground and triplet excited states of most molecules. Consequently collision of ground state $^3\Sigma$-oxygen with an excited triplet species will readily effect population of the oxygen excited singlet states.

Molecular oxygen can insert itself into a substrate by two distinct routes. The first, Type I, involves reaction of ground state $^3\Sigma$-oxygen with a radical as illustrated for sensitised hydrogen abstraction from an alcohol. The product of the reaction is a hydroxyhydroperoxide which is sensitive to moisture and decomposes to a ketone and hydrogen peroxide. The second route, Type II, for oxygen insertion involves reaction of $^1\Delta$-oxygen with a ground state substrate molecule and results in the formation of a peroxide, hydroperoxide or dioxetane.[1] The formation of peroxides (endoperoxides) and $\alpha\beta$-unsaturated hydroperoxides (the Schenck reaction) from Type II

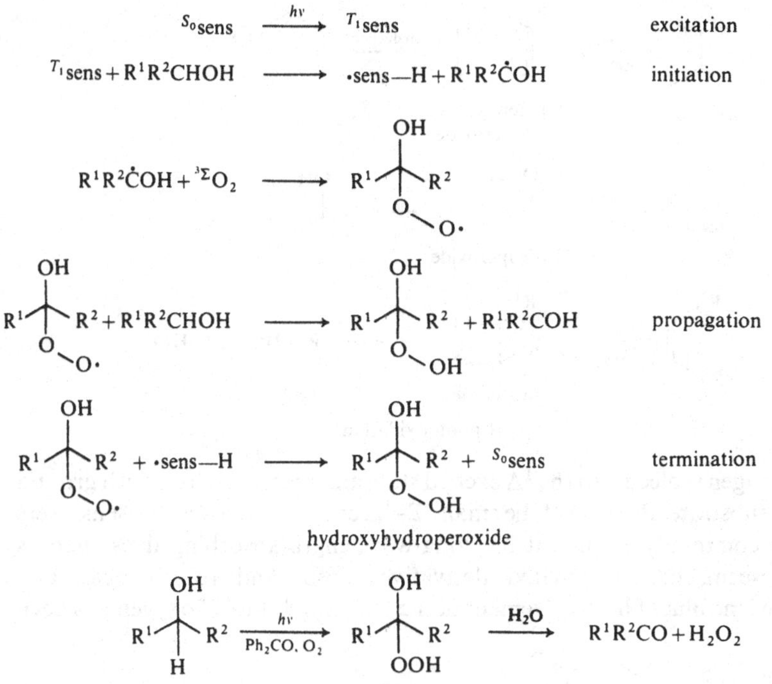

$$S_0\text{sens} \xrightarrow{h\nu} T_1\text{sens} \qquad \text{excitation}$$

$$T_1\text{sens} + R^1R^2CHOH \longrightarrow \cdot\text{sens}-H + R^1R^2\dot{C}OH \qquad \text{initiation}$$

$$R^1R^2\dot{C}OH + {}^3\Sigma O_2 \longrightarrow R^1\!\!\underset{\underset{O\diagdown O\cdot}{}}{\overset{OH}{\diagup\diagdown}}\!\!R^2$$

$$R^1\!\!\underset{\underset{O\diagdown O\cdot}{}}{\overset{OH}{\diagup\diagdown}}\!\!R^2 + R^1R^2CHOH \longrightarrow R^1\!\!\underset{\underset{O\diagdown OH}{}}{\overset{OH}{\diagup\diagdown}}\!\!R^2 + R^1R^2COH \qquad \text{propagation}$$

$$R^1\!\!\underset{\underset{O\diagdown O\cdot}{}}{\overset{OH}{\diagup\diagdown}}\!\!R^2 + \cdot\text{sens}-H \longrightarrow R^1\!\!\underset{\underset{O\diagdown OH}{}}{\overset{OH}{\diagup\diagdown}}\!\!R^2 + S_0\text{sens} \qquad \text{termination}$$

hydroxyhydroperoxide

$$R^1\!\!\underset{\underset{H}{}}{\overset{OH}{\diagup\diagdown}}\!\!R^2 \xrightarrow[Ph_2CO,\,O_2]{h\nu} R^1\!\!\underset{\underset{OOH}{}}{\overset{OH}{\diagup\diagdown}}\!\!R^2 \xrightarrow{H_2O} R^1R^2CO + H_2O_2$$

Type I photooxidation

photooxidation occurs with a large variety or dienes and alkenes and is of synthetic value.[1,2] These products are readily reduced to the corresponding 1,4-diols and allylic alcohols. The formation, and subsequent decomposition, of dioxetanes have only recently been discovered.

The course of a Type II photooxidation reaction involves sensitisation of

Fig. 5.1. Electronic states of the oxygen molecule.

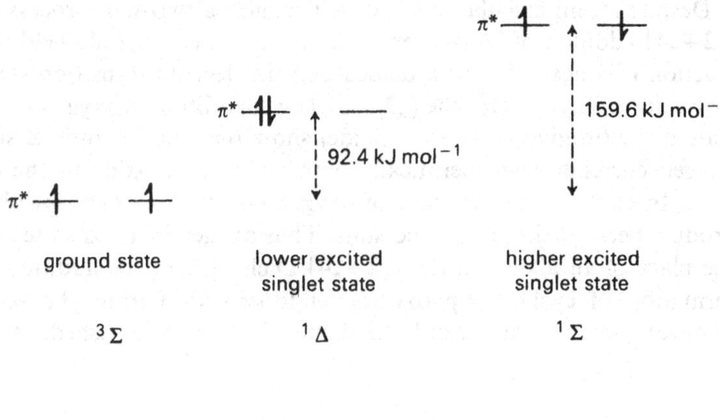

ground state	lower excited singlet state	higher excited singlet state
$^3\Sigma$	$^1\Delta$	$^1\Sigma$

$92.4\ \text{kJ mol}^{-1}$

$159.6\ \text{kJ mol}^{-1}$

perepoxide
(endoperoxide)

hydroperoxide

dioxetane

Type II photooxidation

the oxygen molecule to the $^1\Delta$ excited state and reaction of this with ground state substrate. Because of the small $^3\Sigma$–$^1\Delta$ energy difference, the sensitisers most commonly employed are long-wavelength-absorbing dyes such as fluorescein, its halogenated derivatives eosin and rose bengal, and methylene blue. The involvement of the $^1\Delta$ excited state of oxygen has been

fluorescein

methylene blue

unequivocally established from the identity of the products of such sensitised photooxygenations with those from reaction with $^1\Delta$-oxygen generated from hypochlorite and hydrogen peroxide.[1]

Despite recent calculations[3] which suggest a two-step process for the $[_\pi 2 + _\pi 4]$ addition of $^1\Delta$-oxygen to 1,3-dienes, it is generally held that the reaction is concerted with a delocalised six-electron transition structure. Correlation diagrams for the $[_\pi 2_s + _\pi 4_s]$ cycloaddition of oxygen to ground state diene (to give an endoperoxide) show that the $^3\Sigma$ and $^1\Sigma$ states of oxygen correlate endothermically with product peroxide in the excited state. In contrast the $^1\Delta$ state of oxygen correlates exothermically with product peroxide in the ground state. Thus oxygen in its $^1\Delta$ state can take the place of an alkene in the $[_\pi 2 + _\pi 4]$ Diels–Alder cycloaddition.[4] The formation of cyclic 1,4-peroxides (endoperoxides) from the sensitised photooxygenation of cisoid 1,3-dienes is well established. The sole

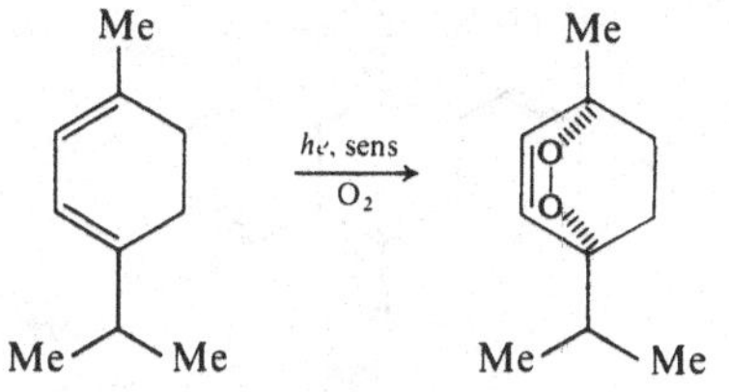

(1)

naturally occurring endoperoxide is ascaridole (1) and this can be synthesised from α-terpinene.

The carbocyclic dienes (2) afford isolable endoperoxides which are easily

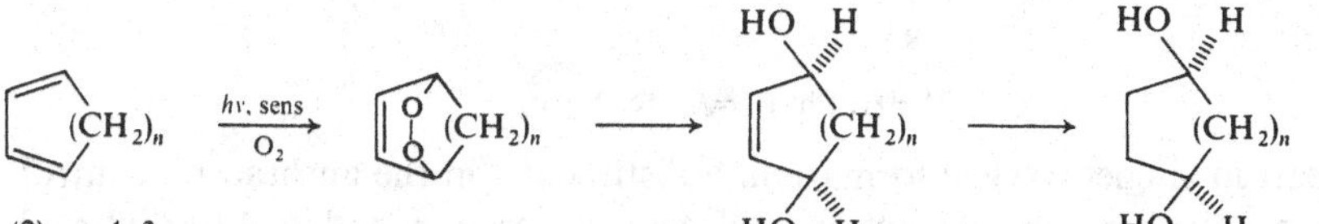

(2) $n = 1$–3

reduced to unsaturated or saturated *cis*-1,4-diols. Cycloaddition of oxygen is not confined to monocyclic cisoid dienes; bicyclohexenyl undergoes cycloaddition to the tricyclic peroxide (3) but reactions of this type are less

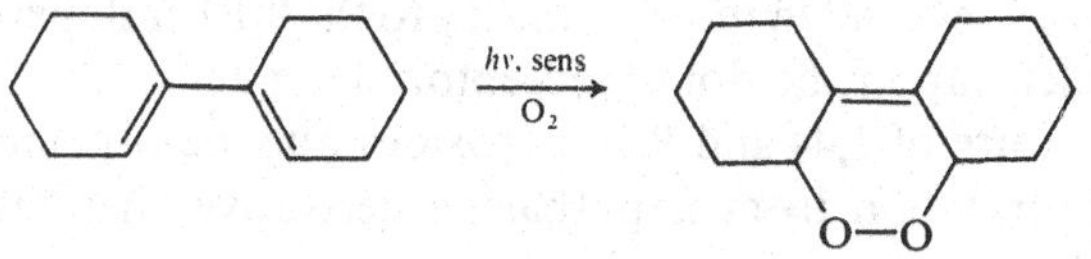

(3)

common. Steroidal dienes undergo stereospecific photooxidation, the endoperoxide being formed by addition of $^1\Delta$-oxygen to the least hindered face of the molecule.

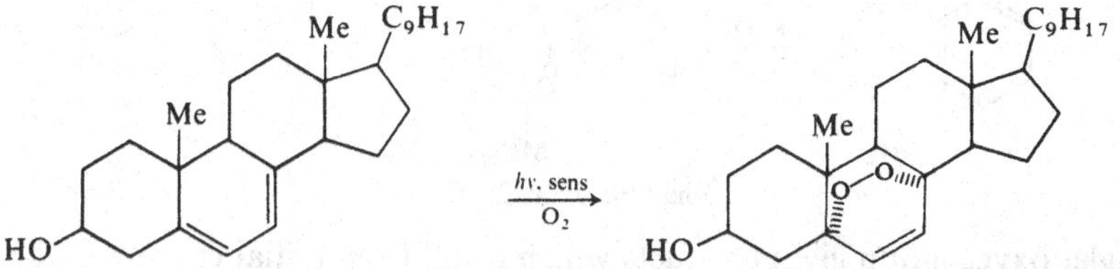

Aromatic systems can undergo direct photooxygenation; anthracene, for example, gives the 9,10-endoperoxide without added sensitiser.[1] Kinetic studies have shown that in this, and other condensed aromatics, the hydrocarbon acts as sensitiser for singlet oxygen formation ($^3\Sigma$–$^1\Delta$) and this, once formed, adds to a second (ground state) molecule of the aromatic compound. However, the aromatic hydrocarbon sensitiser differs from its dye equivalent in that both the S_1 and T_1 states of the hydrocarbon can take

R = H, Ph, Cl

R^1 = H, Ph; R^2 = OMe, NMe_2

part in singlet oxygen formation. Substituents on the anthracene control the regiochemistry of reaction, and the hydrocarbon, and its 9,10-diphenyl and dichloro derivatives afford 9,10-endoperoxide regiospecifically. The presence of electron donating substituents, e.g. OMe, NMe_2, at C_1 and C_4 cause a change in the sites of reaction and only the 1,4-endoperoxide ensues. If the electron donating substituents are located at the C_2- and C_3- positions, however, a reversal in regiochemistry to the 9,10-endoperoxide is observed. Dependent upon the donor–acceptor character of the 1,4,9,10-substituents, a mixture of 1,4- and 9,10-peroxides may be obtained.[1]

Endoperoxide formation from naphthalene derivatives has also been observed but the products have limited stability. Many of these endoperoxides undergo retro-Diels–Alder reaction even at ambient temperature. With polymethoxybenzenes and N,N-dimethylanilines,

half-life 5 h *c.* 25 °C

singlet oxygenation gives products which result from initial endoperoxide formation; no benzenoid peroxide has yet been isolated. The six-electron heteroaromatic analogues are similar.[1] For example, photooxygenation of furan gives the ozonide (4) which decomposes explosively at − 10 °C; in the presence of triphenylphosphine butendial is formed. The peroxides from aryl substituted furans often have enhanced stability with decomposition occurring at ambient temperatures. The sensitivity of pyrroles to air and light is well known and while endoperoxide formation is frequently invoked as the first step of oxygenation, the peroxides have yet to be isolated. An analogous situation pertains to thiophene oxygenations. Thiophene itself is

unreactive to singlet oxygenation but a variety of derivatives give products probably by way of an endoperoxide.

Non-conjugated alkenes and acyclic polyenes readily afford hydroperoxides by Type II photooxidation.[1,4,5] These are easily reduced to allylic alcohols, usually by a reductive work-up. The reaction is characterised by double-bond migration and is stereoselective, the carbon–hydrogen bond broken being perpendicular to the plane of the double bond and on the same face of the molecule as oxygen attack. This is demon-

(5) *a*: $R^1 = D$, $R^2 = H$
   b: $R^1 = H$, $R^2 = D$

(1) *hv*, sens, O_2
(2) reduction

(7)

(6) *a*: $R^2 = H$ (91.5%)
   b: $R^2 = D$ (95%)

strated for the deuterated cholesterols (5*a,b*) where photooxidation, followed by reduction of the hydroperoxide gives (6*a,b*) in which the 7α-hydrogen is lost. Addition of oxygen from above the molecule to give (7) does not occur owing to steric hindrance from the C_{10} angular methyl group. Similarly the *cis*-isomer of the $^2\Delta$-carene, shown in scheme 5.1, gives a sole hydroperoxide whereas the *trans*-isomer with a hindered allylic hydrogen does not react.

Apart from high stereoselectivity the reaction also demonstrates a high and unusual regioselectivity.[8] With trisubstituted alkenes proton abstraction is always from the more crowded side of the double bond, but it

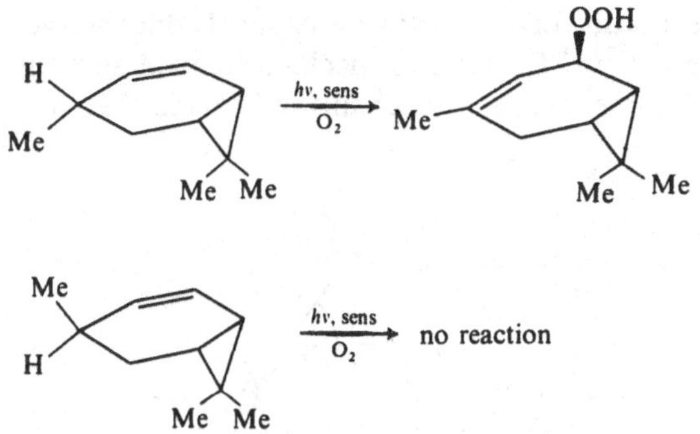

Scheme 5.1

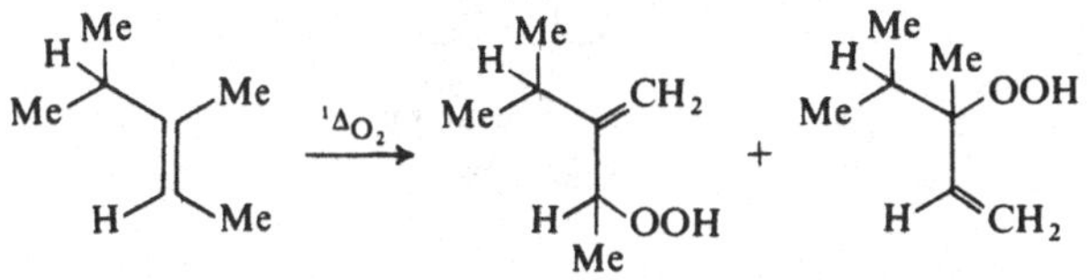

(8)   (9)   (10)
             ~1   :   1

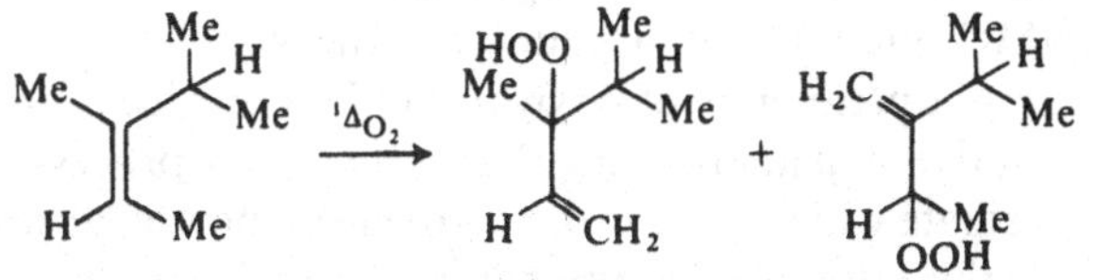

(11)   ~4   :   1

does not involve methine hydrogen atoms. Thus the reaction of alkene (8) with $^1\Delta$-oxygen yields approximately equal amounts of (9) and (10) resulting from hydrogen abstraction from the *cis*-methyl functionalities. In like manner compound (11) suffers proton abstraction predominantly from C_1 *cis* with respect to the isopropyl group. In neither of these examples is any methine proton abstraction from the isopropyl group recorded. This type of regioselectivity is also apparent in cyclic systems as shown for 1-methylcyclobutene.

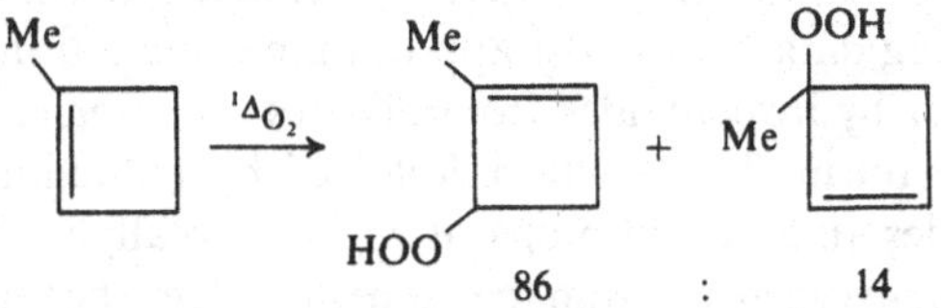

86   :   14

These hydroperoxide forming reactions proceed with very low activation energies, usually below $20\,kJ\,mol^{-1}$, and consequently many of the traditional ideas concerning mechanism are difficult to apply. Nevertheless, any mechanistic explanation of the reaction must satisfy the observed regio- and stereo-selectivities.[1,4,5] In fact two mechanistic explanations are in vogue. The first of these, the ene mechanism (scheme 5.2) involves a

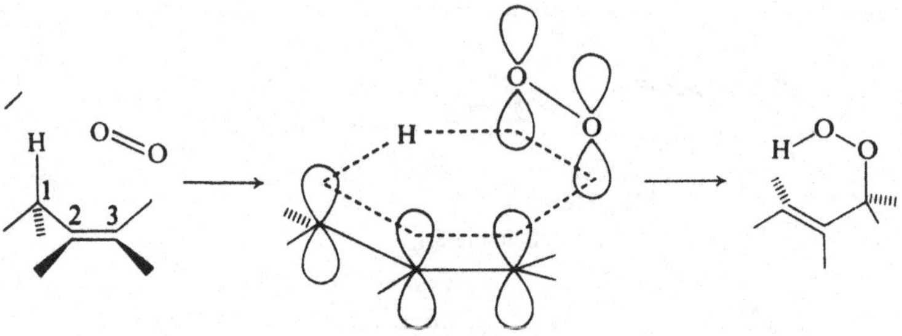

ene mechanism

Scheme 5.2

concerted transformation of substrates into product. If the alkene is regarded as an electron donor and $^1\Delta$-oxygen as an electron acceptor, then the HOMO–LUMO interactions account for the key features of the reaction including the preference of oxygen for the crowded side of the alkene. The alternative explanation involves a two-step process with a perepoxide intermediate which gives hydroperoxide on proton abstraction (scheme 5.3). While this latter mechanism has gained some support it is not

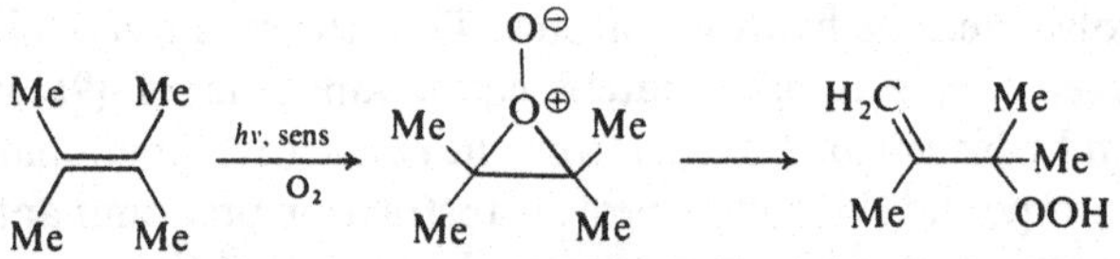

perepoxide

Scheme 5.3

clear why the initial interactions should lead to a perepoxide. In the absence of additional supporting data[6] the most appropriate explanation for the concerted formation of hydroperoxides from alkenes and singlet oxygen must be the ene mechanism. The reaction, followed by reduction of the hydroperoxide, provides an excellent preparative route to allylic alcohols.

The third mode of $^1\Delta$-oxygen addition to alkenes involves the formation of dioxetanes, as demonstrated by the photolysis of *cis-* or *trans-*diethoxyethene at low temperature in the presence of oxygen.[1] The addition is stereospecific and only proceeds with activated alkenes which do not have acidic allylic hydrogen atoms. Orbital symmetry correlations show that the concerted addition of $^1\Delta$-oxygen to a ground state alkene is an allowed $[_\pi 2_a + _\pi 2_s]$ process, providing the ionisation potential of the alkene is low. On warming in benzene, dioxetanes decompose

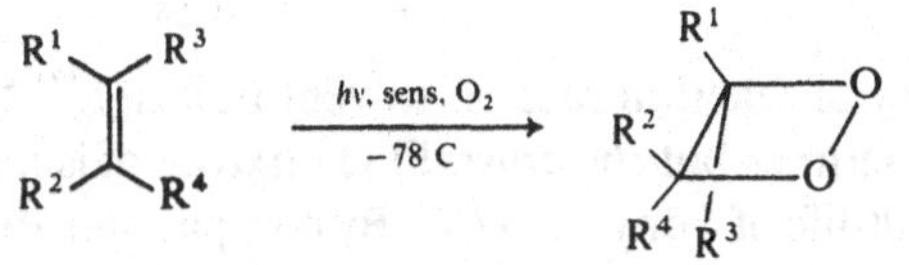

a: $R^1 = R^2 = H$; $R^3 = R^4 = OEt$
b: $R^1 = R^4 = H$; $R^2 = R^3 = OEt$

dioxetane

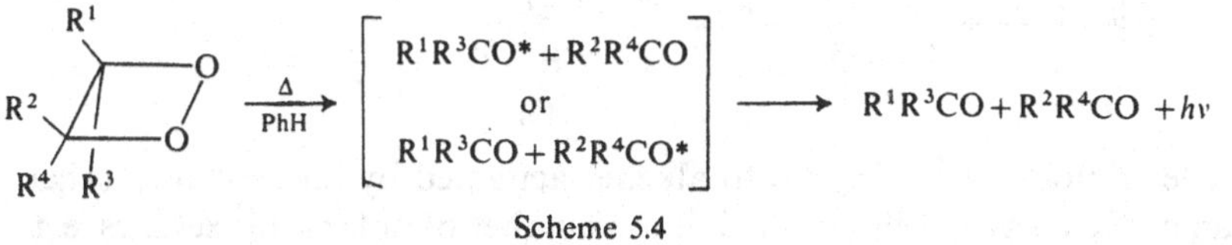

Scheme 5.4

exothermically to give aldehyde and/or ketone products (scheme 5.4). Because an excited state and a ground state molecule give rise to the dioxetane, the symmetry-allowed $[_\sigma 2 + _\sigma 2]$ cycloreversion is accompanied by a bluish luminescence (λ_{max} 430–440 nm). One of the two product molecules resulting from the exothermic decomposition is formed in an electronically excited state. The emission of radiation occurs by radiative deactivation of the excited molecule. Reactions in which luminescence is observed are termed *chemiluminescent reactions*[7] and further examples include the exothermic decomposition of certain acene endoperoxides, e.g. (12). Here the driving force is presumed to be the reestablishment of aromatic character.

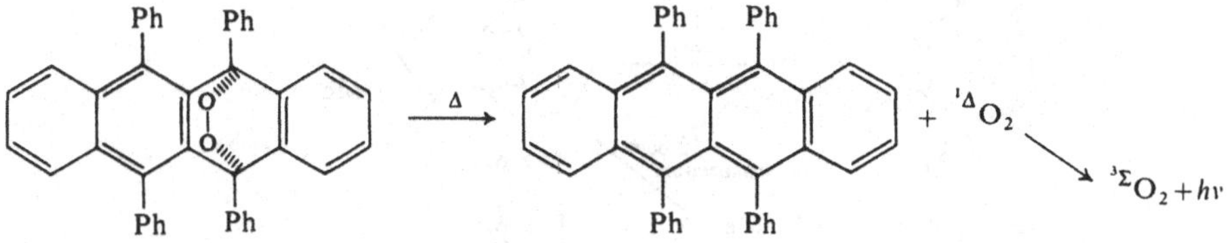

(12)

The formation of dioxetanes involves a more polar transition structure than either endoperoxide formation or the 'ene' reaction as demonstrated

(13)   (14) $\tau^{37\ ^\circ C}$ 4 days

by the dependency of reaction rate on solvent polarity.[8] Most dioxetanes have only short lifetimes but the crystalline mixture of isomers (14) derived from (13) has a half-life of 4 days at 37°C. By comparison attempts to isolate the parent system by photooxygenation of ethene have failed; the compound has become available, however, by dehydrohalogenation[9] of the halohydroperoxide (15).

The addition of $^1\Delta$-oxygen to alkenes activated by functionality other than oxygen is also well known, but the number of isolable dioxetanes, e.g. (16), is not great. The effect of the nitrogen substituent is to alter the

(16)   (17)

mechanism of the reaction and initial electron transfer gives a pair of radical ions which collapse to product.[10] Thioethenes also react with singlet oxygen probably by way of an initially generated dioxetane. Thus tetramethylthioethene gives dimethyldithiooxalate (17) together with dimethyldisulphide. The first isolable dioxetane from a bis-alkylthioethene has only recently been recorded.[11] When allowed to warm to ambient temperature the compound undergoes bond cleavage in two separate ways. The anticipated carbon–carbon and oxygen–oxygen cleavage provide ring-expanded product while carbon–sulphur and oxygen–oxygen fission results in the cycloheptandione.

5.2   Oxidative coupling[12]

The cyclisation of 1,2-diarylethenes under oxidative conditions to give the corresponding phenanthrene derivative is perhaps the best known of oxidative photocoupling reactions.[12] If a *trans*-alkene is employed an initial *trans–cis* photoisomerisation is necessary in order to bring the aromatic rings within bonding distance and hence it is preferable for the preparation of phenanthrenes to start with *cis*-alkenes.

The reaction proceeds by six-electron conrotatory electrocyclic closure across the hexatriene component of the *cis*-stilbene moiety to yield a *trans*-dihydrophenanthrene (18) (section 2.2). The ring closure is a singlet state

(18)

concerted reaction since triplet sensitisers do not sensitise the reaction and triplet quenchers are without effect. Moreover, the reaction does not proceed when the stilbene carries substituents such as acetyl and nitro which enhance intersystem crossing (ISC). The involvement of a dihydrophenanthrene intermediate in the reaction is well established and the lifetimes of these species range from a few tenths of a second to a period of days. *cis*-Diphenylethene produces the highly conjugated 4a,4b-dihydrophenanthrene (18) which is yellow (λ_{max} 447 nm) and stable under anaerobic conditions. This substrate (18) cleaves thermally (non-concerted) and photochemically to regenerate starting material while in the presence of oxygen, or other suitable oxidants, dehydrogenation occurs to give phenanthrene. The stereochemistry at C_{4a} and C_{4b} of the intermediate cyclised species (18) has been shown to be *trans* from the isolation of a stable analogue in the photolysis of 3,4-di(4-hydroxyphenyl)hex-3-ene. The dienol (19) is the tautomer of diketone (20) to which it owes its stability. The 4a,4b hydrogens of (20) resonate as a singlet in the n.m.r. spectrum at 2.3 p.p.m., and molecular hydrogen is not lost from the molecule to any significant extent in the mass spectrometer which is consistent with a *trans*-hydrogen configuration.

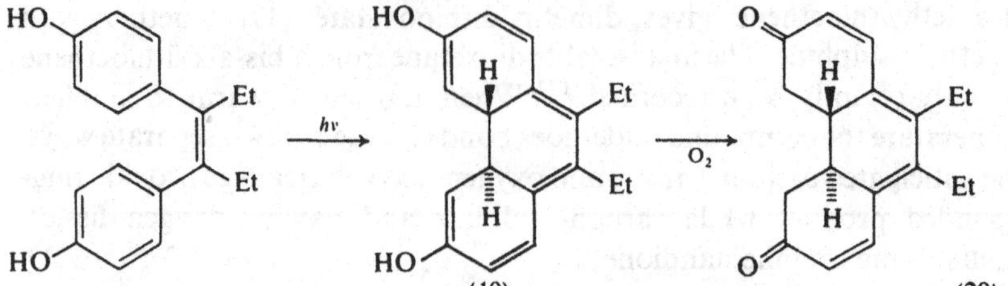

(19)   (20)

The reaction sequence has found application in the synthesis of a large number of polycyclic and polyheterocyclic aromatic compounds, certain alkaloids, and other substituted ring systems. For example, picene (21) is formed from 1,2-distyrylbezene; triphenylene (22) from *ortho*-terphenyl;

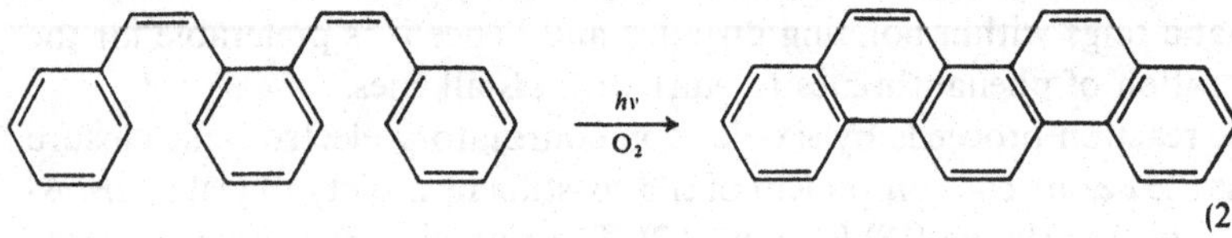

(21)

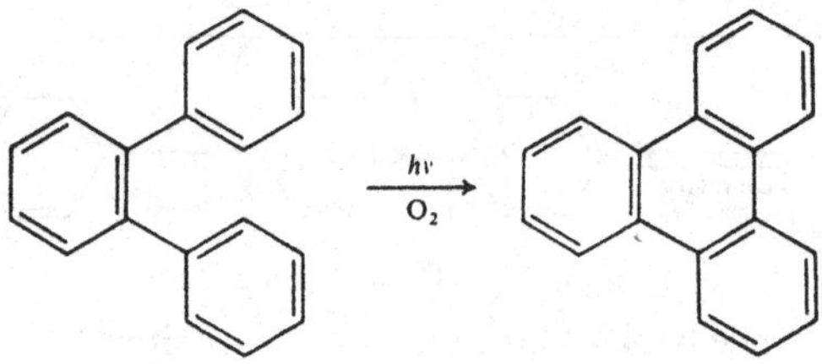

(22)

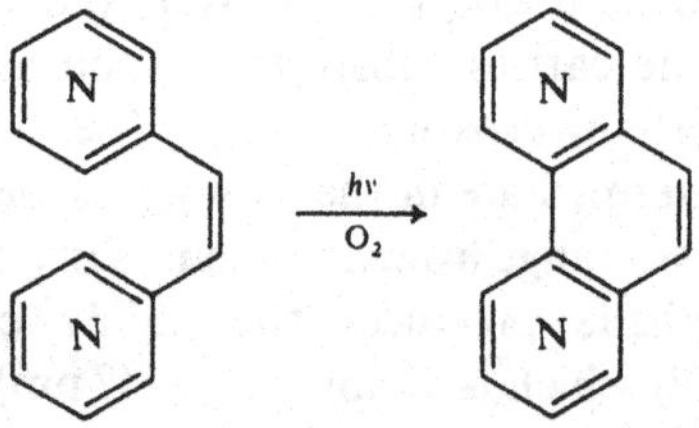

(23)

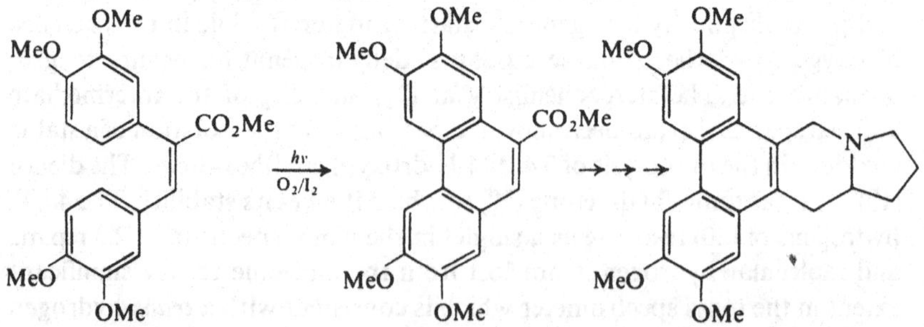

(24)   ($\pm$) (25)

phenanthrolines (23) (the symbol indicates the presence in the structure of pyridinic nitrogen atoms) from 1,2-dipyridylethenes; (±)-tylophorine (25) from the stilbene ester (24).

In many reactions iodine has been used as the oxidant and cases of oxidation by copper halide are known. There are several variants to these photocyclisations. For example, irradiation of either *ortho-* or *para-*iodostilbene leads to phenanthrene. The product results from the ready photolytic cleavage of a carbon–iodine bond. One of the most striking features of the formation of polycyclic aromatic hydrocarbons by photocyclodehydrogenation is the regioselectivity of the reactions. For example, photolysis of (26) yields hexahelicene (27) but not its isomer (28).

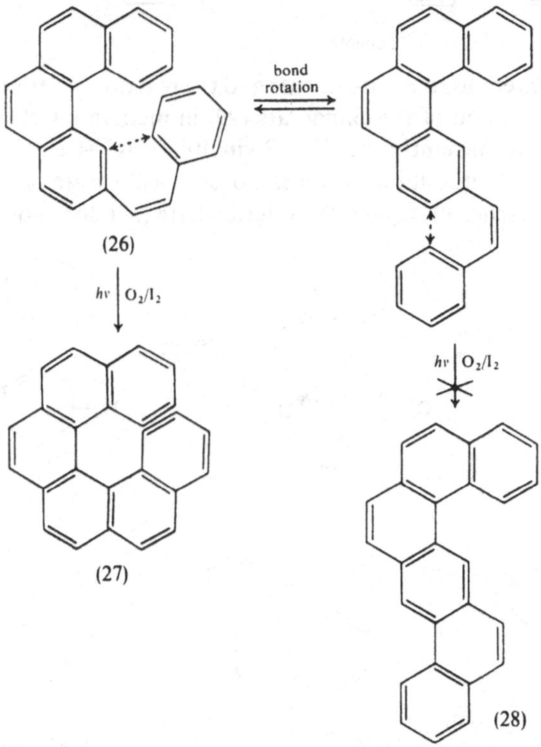

Reactivity parameters have been devised for the photocyclisation of 1,2-diarylethenes. However, the conformational preference of the alkene, the relative lifetimes of the possible dihydrophenanthrenes, the preference for formation of planar versus non-planar polycyclic aromatic products, and the competition of *cis–trans* photoisomerisation complicate predictions for a given reaction.[12]

A different outcome of the photocyclisation reaction occurs when the

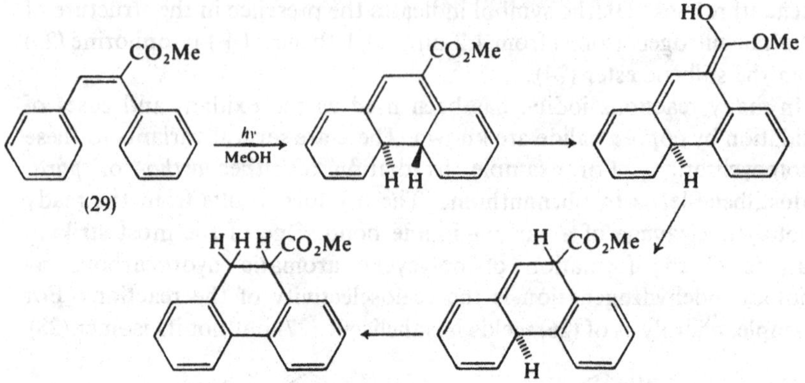

Scheme 5.5

diarylethene carries enolisable substituents and the reaction is performed in a protic solvent; reaction of the cinnamate (29) in methanol (scheme 5.5) gives a 9,10-dihydrophenanthrene. If a 2-vinylbiphenyl is employed as substrate, cyclisation without oxidation can occur as illustrated in scheme 5.6. Even in the presence of oxygen the intermediate (30) does not oxidise

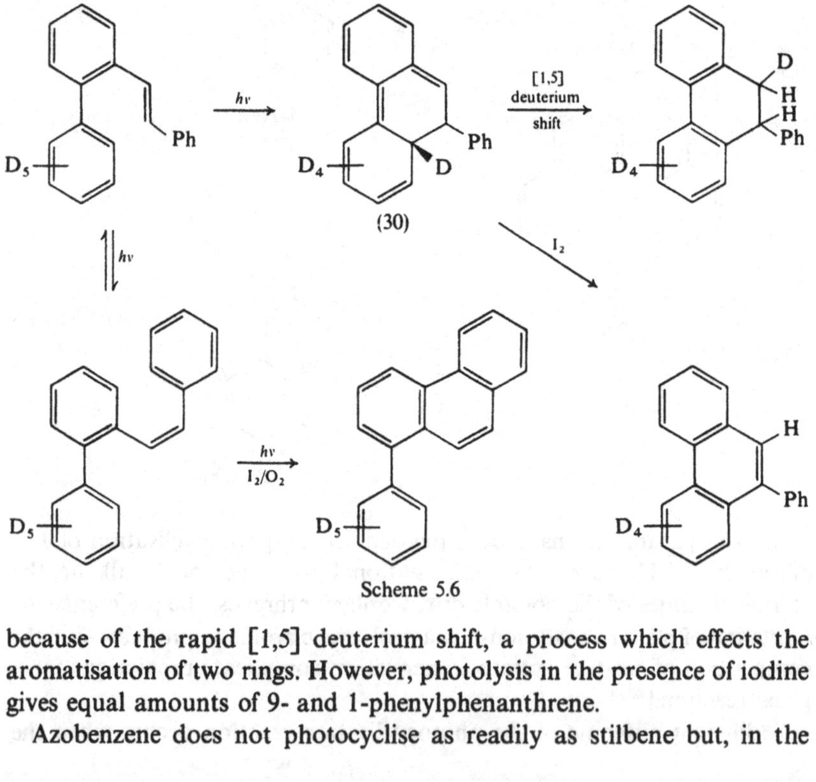

Scheme 5.6

because of the rapid [1,5] deuterium shift, a process which effects the aromatisation of two rings. However, photolysis in the presence of iodine gives equal amounts of 9- and 1-phenylphenanthrene.

Azobenzene does not photocyclise as readily as stilbene but, in the

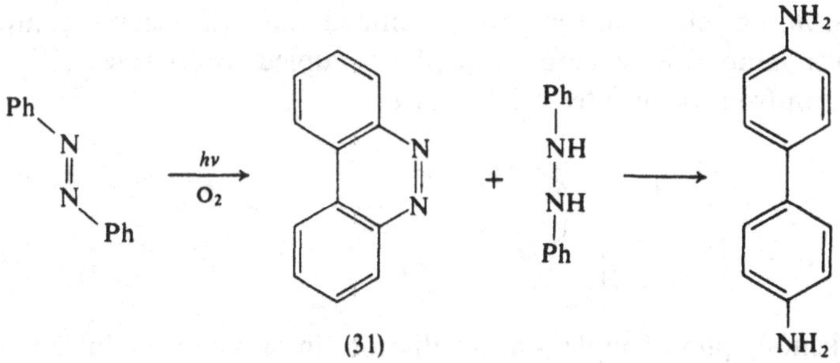

(31)

presence of sulphuric acid, benzidene and benzo[*c*]cinnoline (31) are formed. The reaction is a disproportionation reaction; the hydrogen lost in

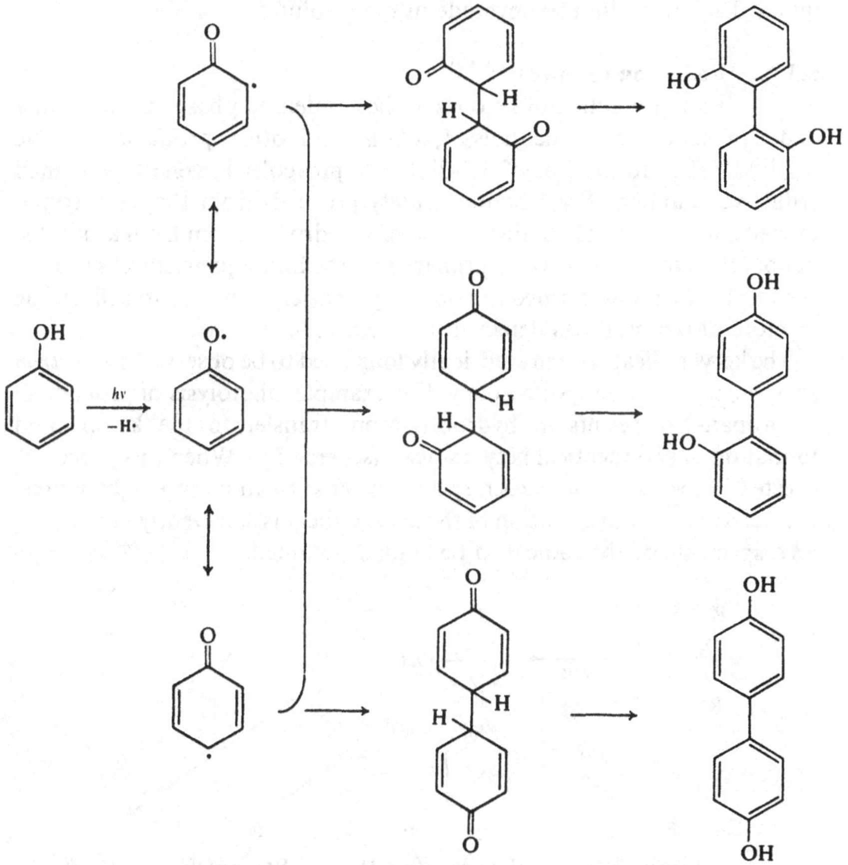

Scheme 5.7

the cyclisation reduces a molecule of azobenzene to hydrazobenzene which subsequently rearranges to benzidene under the acidic conditions of the reaction. Cyclisation is not confined to 1,2-diaryl compounds; diphenylamines, for example, give photocoupled products which arise from reaction from the excited triplet state.

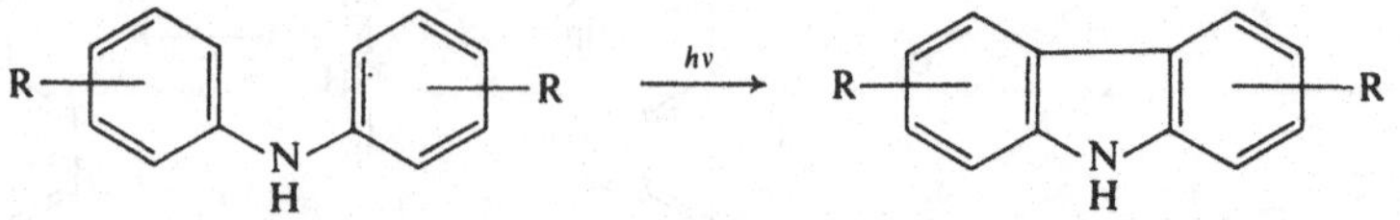

Phenols photocouple on irradiation in aqueous solution, to give mixtures of biphenols. The suggested mechanism involves the formation of phenoxyl radicals which can couple at the *ortho-* and *para*-positions (scheme 5.7). The reaction is analogous to phenolic radical coupling initiated by potassium ferricyanide in basic solution.

5.3   Reduction reactions[13-19]

Photoreduction of ketones in the condensed phase occurs readily and can give rise to alcohols, 1,2-diols, and other products (scheme 5.8).[13,14] The formation of 1,2-diols (or pinacols) is generally termed reductive coupling. The reaction usually proceeds from the n–π* triplet excited state and involves abstraction of a hydrogen atom from a suitable donor (often the solvent) as the primary process. Subsequent stabilisation of the ketyl radical thus formed can occur by a variety of means to deliver one or more of the products depicted in scheme 5.8.

The ketyl radical is often sufficiently long lived to be observed by *electron spin resonance* (e.s.r.) spectroscopy. For example, photolysis of propanone in propan-2-ol results in hydrogen atom transfer to the ketone and formation of two identical ketyl radicals (scheme 5.9). When this process is effected in the cavity of an e.s.r. spectrometer with an intense light source, the steady state concentration of the ketyl radical is sufficiently high for an e.s.r. spectrum of the radical to be readily obtained.

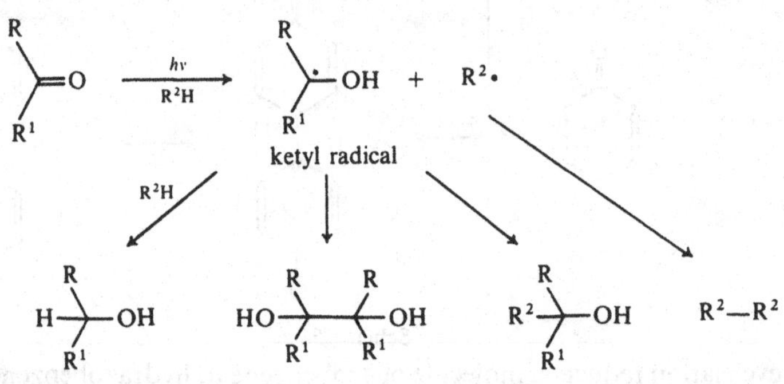

Scheme 5.8

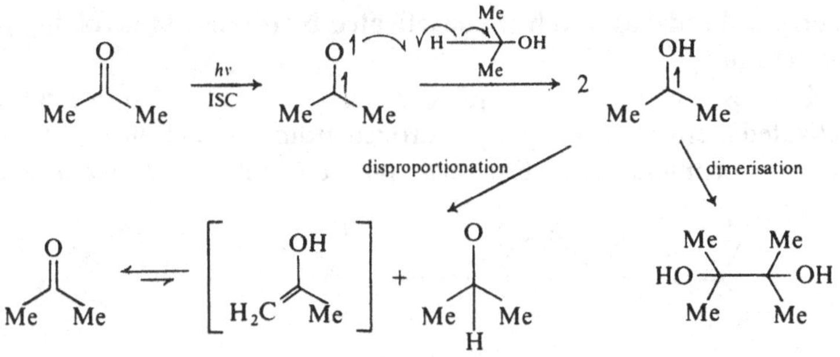

Scheme 5.9

The excited carbonyl group is essentially diradical in nature and abstracts hydrogen from C_2 of propan-2-ol as does an alkoxy radical. The preference for C_2 hydrogen abstraction over abstraction of the hydroxyl hydrogen can be understood from the bond dissociation energies. The former bond has a lower bond dissociation energy ($c.$ 370 kJ mol^{-1}) than does the oxygen–hydrogen bond ($c.$ 415 kJ mol^{-1}). The carbon–hydrogen bond strength of the hydrogen donor and the energy of the carbonyl excited state are particularly important in determining the feasibility of a given hydrogen abstraction reaction. The higher the excited state energy of the carbonyl and the lower the bond strength of the hydrogen donor the more likely the reaction is to occur.

The ketyl radical which is produced when an excited carbonyl group abstracts a hydrogen atom can dimerise. For photolysis of propanone in propan-2-ol, 2,3-dimethylbutan-2,3-diol (pinacol) is formed as the sole product of reaction (scheme 5.9). The quantum yield for product formation and the quantum yield for the disappearance of ketone are lower than might be expected since disporportionation of the ketyl radicals competes with diol formation. If the reduction is carried out with cyclohexane as solvent, a large number of products are formed from both the ketyl (isopropyl) and cyclohexyl free radicals, cf. scheme 5.8. Consequently, as a general preparative method for 1,2-diol formation it is best to employ the alcohol derived from the ketone as the reaction solvent.

Reductive coupling occurs for aromatic ketones, benzophenone giving benzpinacol as the sole product when the reaction is carried out in benzhydrol as solvent. If an alcohol that forms a stable radical is used as the hydrogen donor in the reaction, the pinacol derived from the alcohol will be a major product. The most suitable hydrogen donors are those that form

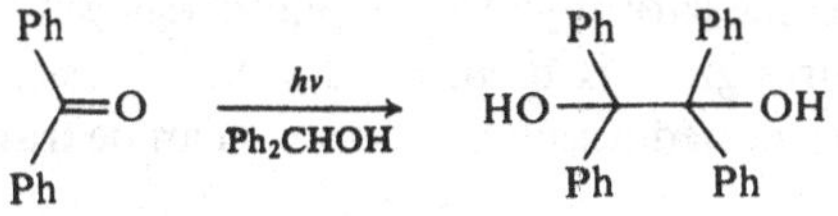

benzpinacol

short-lived radicals which are deactivated by routes not involving pinacol formation.

Ketones, aldehydes and quinones can also add photochemically to activated methylene groups by hydrogen atom abstraction and subsequent radical recombination. For example, an allylic hydrogen atom of

a: $R^1 = R^2 = Me$
b: $R^1 = H, R^2 = Et$

cyclohexene is transferred to the excited carbonyl oxygen atom and the resulting radicals subsequently recombine. Benzylic compounds also provide a source of active hydrogen, irradiation of benzophenone with diphenylmethane giving only the mixed radical coupled product (33a).

$$Ph_2CO + ArCH_2Ar \xrightarrow{h\nu} Ph_2C(OH)C(OH)Ph_2 + Ph_2C(OH)CHAr_2$$
$$(32) \qquad\qquad (33)$$

a: $Ar = C_6H_5-$
b: $Ar = p\text{-}MeOC_6H_4-$

$$+ Ar_2CHCHAr_2$$
$$(34)$$

With bis-(4-methoxyphenyl)methane all possible radical coupled products (32b)–(34b) are formed. With toluene and 9,10-phenanthraquinone, an ether and a ketol result from radical coupling. For *ortho*-alkyl aryl ketones

ketol   ether

with a benzylic hydrogen γ to the carbonyl group, photolysis can result in internal hydrogen transfer through a six-membered cyclic transition state structure (scheme 5.10). The intermediate enol so formed can be trapped by addition with a dienophile.

Photoreduction generally proceeds from the n–π* triplet state. However,

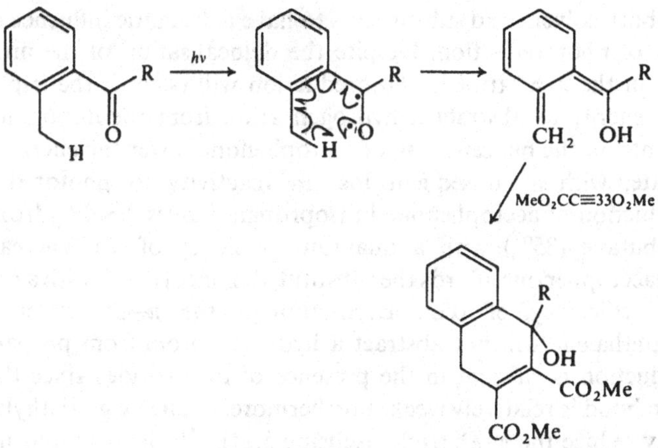

Scheme 5.10

with aromatic ketones both the n–π* and π–π* triplet states can be populated, the former being considerably more reactive and efficient in photoreduction. The n–π* triplet excited state behaves in a manner

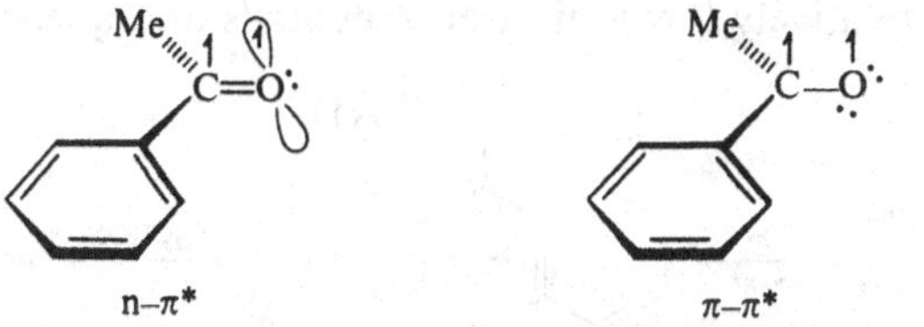

analogous to that of an alkoxy radical with one unpaired electron on the oxygen atom. In the π–π* triplet the unpaired electrons are delocalised over both the carbon and the oxygen atoms. The difference in reactivity of the two states is marked: in isopropanol acetophenone is photoreduced from the n–π* triplet while 2-acetylnaphthalene is not since the lowest excited state of this ketone is the π–π* triplet which has insufficient energy to abstract a hydrogen atom from the solvent.

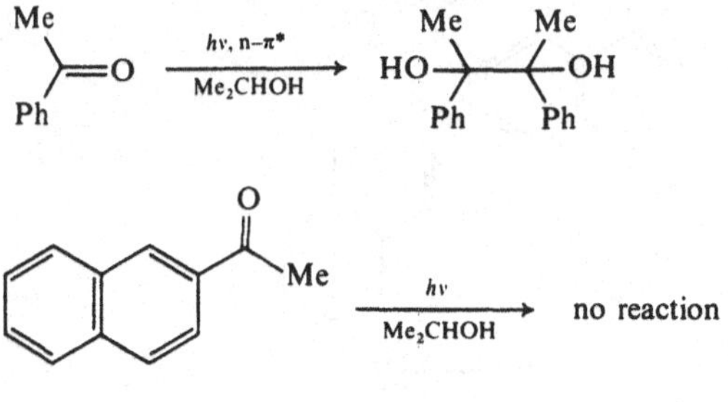

The n–π* and the π–π* excited states of acetophenone are sufficiently close for both solvent and substituents to have a dramatic influence on the efficiency of photoreduction. Despite the delocalisation of the unpaired electrons in the $\pi-\pi^*$ triplet, photoreduction will occur if the triplet has sufficient energy to abstract a hydrogen atom from the donor. Methyl substituents on the benzene ring of acetophenone lower the energy of the $\pi-\pi^*$ state, with a consequent loss in reactivity to photoreduction. Photoreduction of acetophenone in isopropanol gives 2,3-dihydroxy-2,3-diphenylbutane (85%) with a quantum efficiency of 0.4 whereas 3,4-dimethylacetophenone affords the substituted pinacol (83%) with a reduced quantum efficiency of 0.15. Even though the $\pi-\pi^*$ triplet of 2-acetylnaphthalene will not abstract a hydrogen atom from propan-2-ol, photoreduction is efficient in the presence of tin hydrides since the tin–hydrogen bond is relatively weak. Furthermore, amines, e.g. triethylamine, effectively reduce the $\pi-\pi^*$ triplet (scheme 5.11).[15] The reactions involve electron transfer which is favoured when the ionising potential of the solvent is low. Where the ionisation potential is relatively high, as for hydrocarbon and alcohol solvents, quenching of the excited carbonyl occurs by hydrogen abstraction and the formation of ketyl and hydrocarbon radicals. However, amines, sulphides, and unsaturated hydrocarbons have relatively low ionisation potentials and quenching of

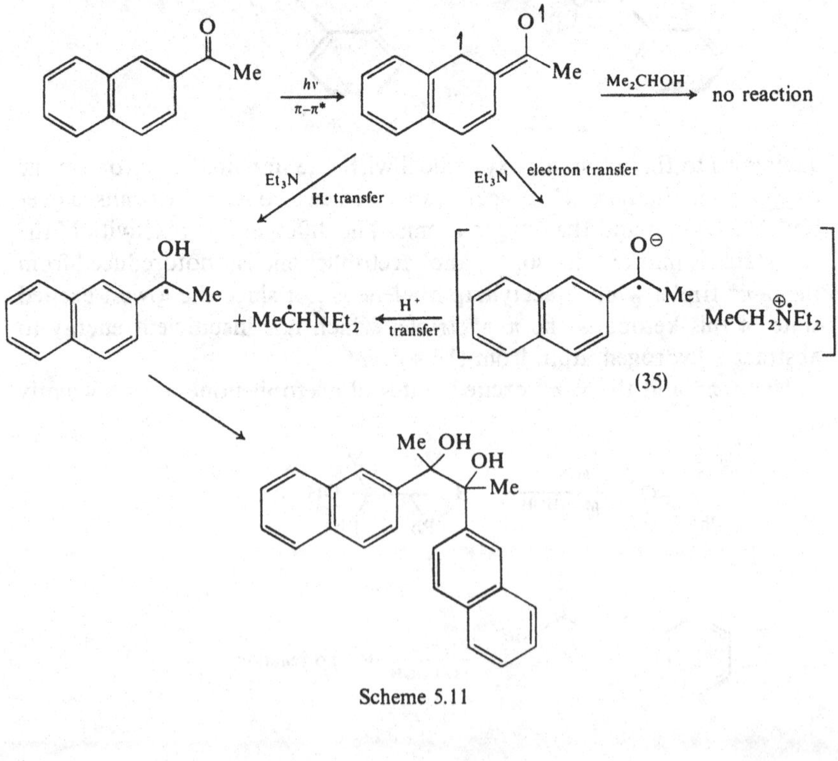

Scheme 5.11

the excited state of a carbonyl can occur via either a charge-transfer interaction or complete electron transfer. Depending upon the strength of the bond to the available hydrogen atom of the donor and the ionisation potential of the donor, *viz.* the ability to donote an electron, there is a continuum of mechanisms between hydrogen transfer and electron transfer (scheme 5.11). Hydrogen transfer results in ketyl and amine radicals, the latter undergoing reaction with a second molecule of excited ketone to produce a further ketyl radical and an imine. Electron transfer processes result in a contact radical ion pair or a charge-transfer exciplex (35) which is converted into a ketyl radical and amine radical on proton transfer (scheme 5.11). This process is also analogous to the quenching of excited benzophenone by aliphatic sulphides. Alkyl sulphides can act as weak reducing agents (path (*b*), scheme 5.12), but the back transfer of an electron in the complex to give ground state ketone (path (*a*), scheme 5.12) is the preferred reaction path.

$$Ph_2CO^* + R^1CH_2SR^2 \underset{(a)}{\rightleftharpoons} [Ph_2\overset{.}{C}-\overset{\ominus}{O} \quad R^1CH_2\overset{\oplus\cdot}{S}R^2]$$

$$\downarrow (b)$$

$$products \longleftarrow Ph_2\overset{.}{C}-OH + R^1\overset{.}{C}HSR^2$$

Scheme 5.12

Electron donating substituents in an aromatic ring also influence photoreduction by assisting in the formation of less active charge-transfer complexes. For example, photoreduction of *para*-aminobenzophenone is not practicable in isopropanol solution because of excitation to the low-lying and chemically inactive charge-transfer triplet. In this instance almost complete electron-transfer occurs from the heteroatom substituent to the carbonyl group. This electron shift reverses the electrophilicity of the carbonyl oxygen atom in the excited state making it nucleophilic. Reduction can be effected, however, by employing triethylamine and the quantum efficiency is greatest when hydrocarbon diluents are employed.

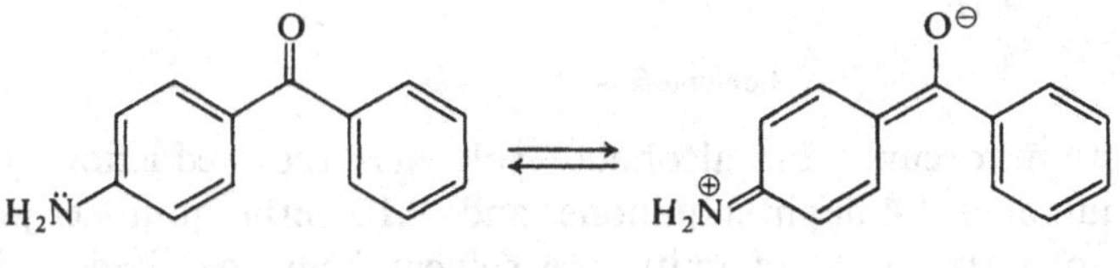

Under conditions where pinacolisation is favoured, $\alpha\beta$-unsaturated ketones are reduced to the corresponding saturated ketone (scheme 5.13). The initially generated and stabilised ketyl radical abstracts a second hydrogen atom from the solvent to achieve the reduction. The

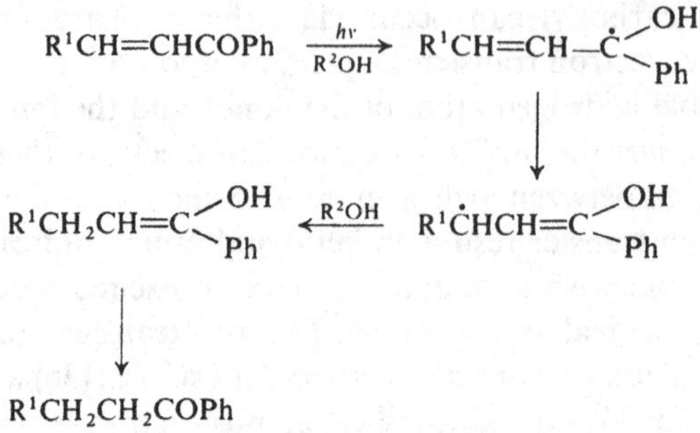

$$R^1CH=CHCOPh \xrightarrow[R^2OH]{hv} R^1CH=CH-\overset{\cdot}{C}\overset{OH}{\underset{Ph}{\diagdown}}$$

$$R^1CH_2CH=C\overset{OH}{\underset{Ph}{\diagdown}} \xleftarrow{R^2OH} R^1\overset{\cdot}{C}HCH=C\overset{OH}{\underset{Ph}{\diagdown}}$$

$$R^1CH_2CH_2COPh$$

Scheme 5.13

photoreduction of aldehydes has found less application than the photoreduction of ketones. Irradiation of benzaldehyde or *para*-methoxybenzaldehyde in ethanol does, however, lead to pinacol formation.

$$ArCHO \xrightarrow{hv}{EtOH} ArCH(OH)CH(OH)Ar$$

The photolysis of quinones in a variety of media results in photoreduction.[16] Hydrogen abstraction from the solvent by the excited quinone results in the formation of the semiquinone radical which subsequently disproportionates thermally to quinone and quinol (scheme 5.14). The efficiency of the reaction depends on the ease by which hydrogen

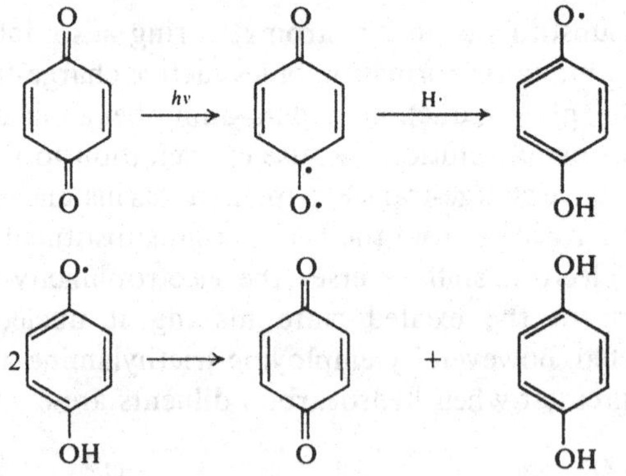

Scheme 5.14

abstraction occurs. In alcoholic solvents the reduction of 1,4-benzoquinone, 1,4-naphthoquinone and 9,10-anthraquinone proceeds with a quantum yield of unity, the solvent being oxidised to ketone. However, alcoholic solvents are generally less effective quenchers than phenols. With 1,2-quinones, photoreduction in alcoholic media results in the corresponding catechol. In hydrocarbon solvents with an available hydrogen atom, however, *ortho*-hydroxyethers can result from radical coupling in the singlet excited state, a process which may be facilitated by a cyclic transition state structure. For coupling from the triplet state, the reaction involves the formation of discrete semiquinone and alkyl radicals. For photoreduction to take place, oxygen must be absent since in the presence of oxygen semiquinone is oxidised to quinone. The existence of the semiquinone radical in these photoreductions has been confirmed by e.s.r. studies.

The reduction products from substituted quinones are very dependent upon the reaction conditions employed. In isopropanol, 2-t-butyl-1,4-benzoquinone yields the quinol and while hydrogen atom transfer from the solvent gives both of the possible semiquinone radicals (36) and (37), the

latter is formed five times faster[17] even though it is thermodynamically the less stable of the pair. In other solvents intramolecular hydrogen transfer from the t-butyl group can dominate and intermediates with the symmetry of a spirocyclopropane are thought to be involved since the reaction gives products with a rearranged side chain (scheme 5.15). The behaviour of 2-amino-1,4-quinones with a γ-hydrogen atom is similar.[18] The reactions involve a Norrish Type II γ-hydrogen atom transfer followed by ring closure and electron demotion to a zwitterionic intermediate, e.g. (38) (scheme 5.15) and (39). With the heteroatom present in the side chain the usual product is that of ring closure and many examples of this reaction are known.[18]

The behaviour of substituted imines on photolysis has been shown to parallel the photoreduction of carbonyl compounds. Irradiation of benzaldehyde N-cyclohexylimine results in the photocoupled reduction product (40). The reaction can be sensitised by benzophenone and the

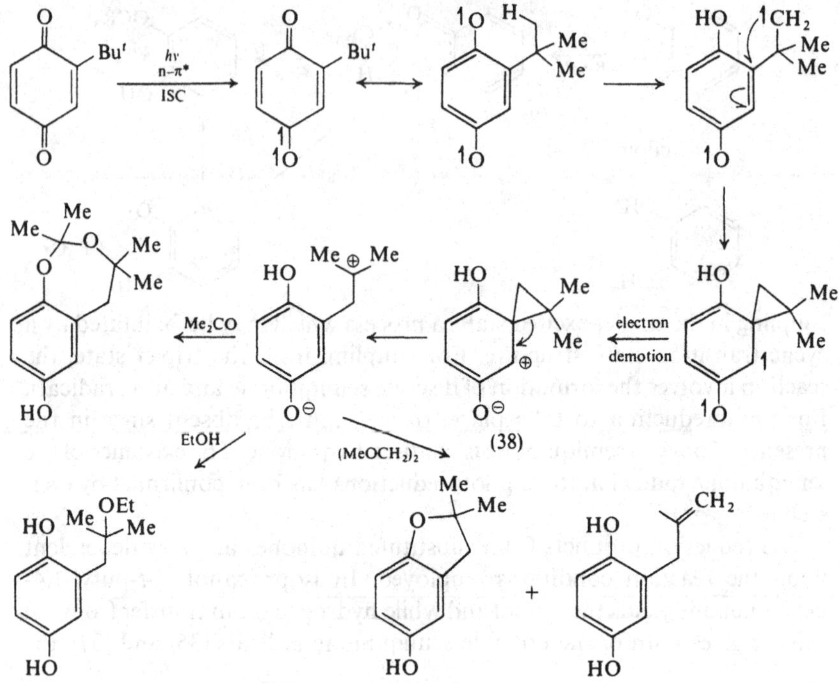

Scheme 5.15

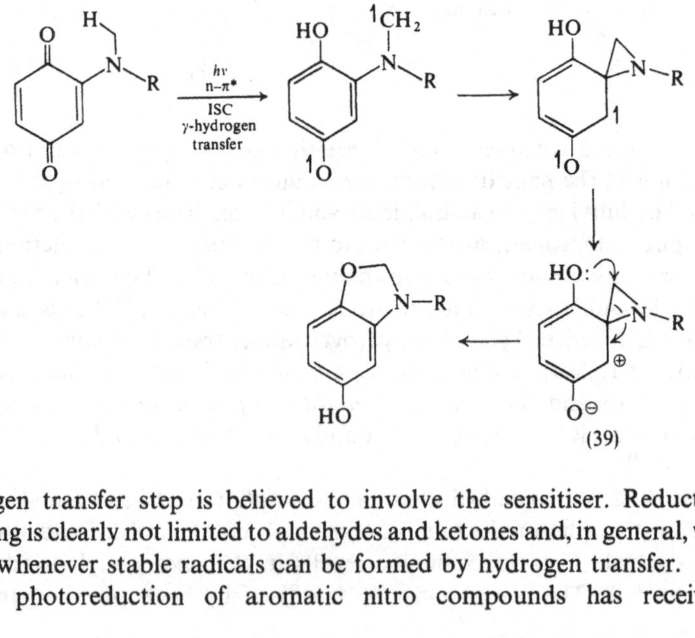

(39)

hydrogen transfer step is believed to involve the sensitiser. Reductive coupling is clearly not limited to aldehydes and ketones and, in general, will occur whenever stable radicals can be formed by hydrogen transfer.

The photoreduction of aromatic nitro compounds has received

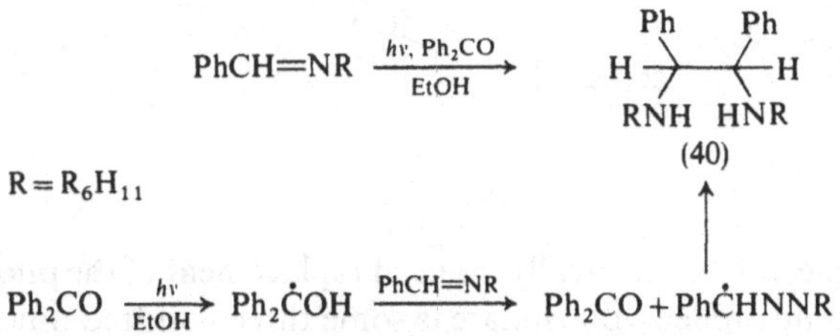

$$PhCH=NR \xrightarrow[\text{EtOH}]{hv,\ Ph_2CO} \quad \underset{RNH\quad HNR}{\overset{Ph\qquad Ph}{H\searrow\!\!\!\diagup\qquad\diagdown\!\!\!\swarrow H}}$$

(40)

$R = R_6H_{11}$

$$Ph_2CO \xrightarrow[\text{EtOH}]{hv} Ph_2\dot{C}OH \xrightarrow{PhCH=NR} Ph_2CO + Ph\dot{C}HNNR$$

considerable attention.[19] The reaction involves the triplet excited state which reacts with solvent to give the nitroso aromatic which is subsequently consumed in yielding arylhydroxylamine as the final product of the reaction (scheme 5.16). With *para*-nitroaniline the lowest energy triplet has

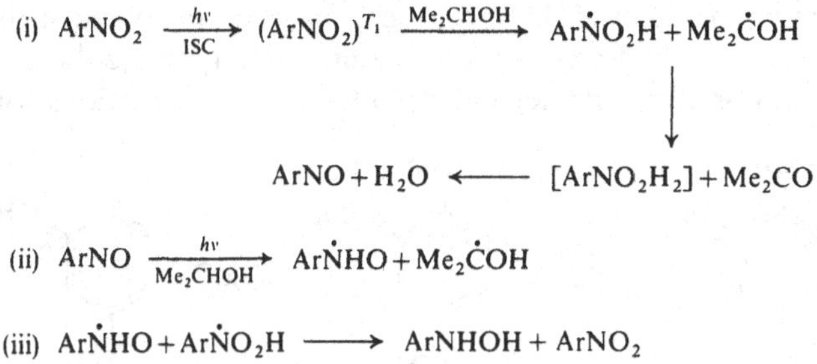

(i) $ArNO_2 \xrightarrow[\text{ISC}]{hv} (ArNO_2)^{T_1} \xrightarrow{Me_2CHOH} Ar\dot{N}O_2H + Me_2\dot{C}OH$

$$ArNO + H_2O \longleftarrow [ArNO_2H_2] + Me_2CO$$

(ii) $ArNO \xrightarrow{hv}{Me_2CHOH} Ar\dot{N}HO + Me_2\dot{C}OH$

(iii) $Ar\dot{N}HO + Ar\dot{N}O_2H \longrightarrow ArNHOH + ArNO_2$

Scheme 5.16

intramolecular charge-transfer character since the reaction is inefficient in isopropanol ($\Phi = 7.2 \times 10^{-4}$) when compared to similar reaction of nitrobenzene ($\Phi = 1.14 \times 10^{-2}$).

5.4 Substitution reactions[20-23].

Many aromatic compounds undergo photosubstitution in the presence of a nucleophilic reagent, but the number of reports of electrophilic aromatic photosubstitution are few. Nucleophilic aromatic photosubstitution reactions often give high yields of product. For heterocyclic and polycyclic aromatic compounds the yields of such products are frequently higher than from simple benzene derivatives.[20]

For aromatic systems substituted with electron-withdrawing groups (e.g. $-NO_2$, $-COMe$, $-CN$), the nucleophile is directed into the *meta*-position while electron donating substituents (e.g. $-OMe$, $-OH$, $-NR_2$, -alkyl) direct the nucleophile to the *ortho*- and *para*-positions. Polycyclic aromatics frequently exhibit 'α-reactivity' while substrates with more than one substituent exhibit a merging (resonance) stabilisation in product formation.

Substituents on an aromatic ring exhibit the reverse effect on orientation, reaction rate and partial rate factors in the excited state than in the ground state. For example, photochemically induced replacement of the phosphate by hydroxyl in 3-nitrophenylphosphate is some three hundred times faster than similar reactions for the *ortho-* and *para*-isomers. This does not unequivocally demonstrate that the nitro group is activating for *meta*-substitution as the effect could result from a longer excited state lifetime for the *meta*-substituted compound than its *ortho-* and *para*-isomers. It is better to compare *meta-* and *para*-substitution within the same molecule and determine the effect of the substituent on reaction course in this way. The substitution of hydroxyl *meta* to the nitro group of 1,2-dimethoxy-4-nitrobenzene (4-nitroveratrole) under photochemical conditions compares

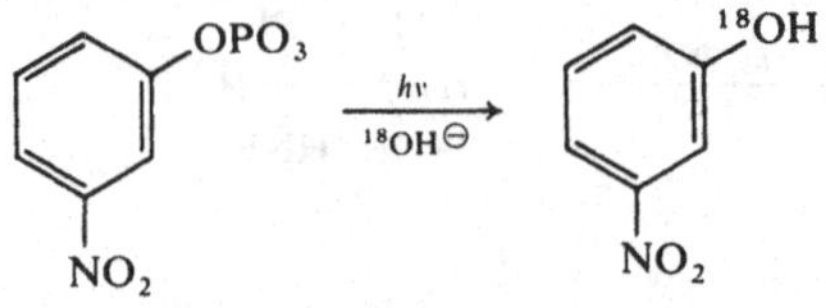

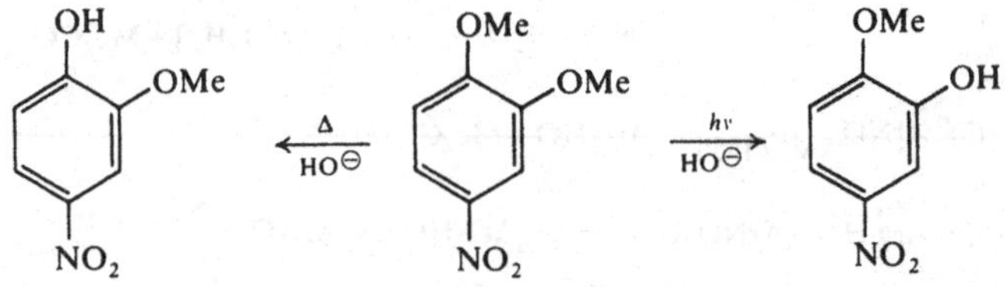

with *para*-substitution of hydroxyl under thermal conditions. Photosubstitution *meta* to the nitro group also occurs when amines and cyanide ions are used as nucleophiles. Photocyanation of the three nitroanisoles illustrates the *meta*-directing influence of the nitro group (scheme 5.17). In the naphthalene derivatives (41) and (42) striking

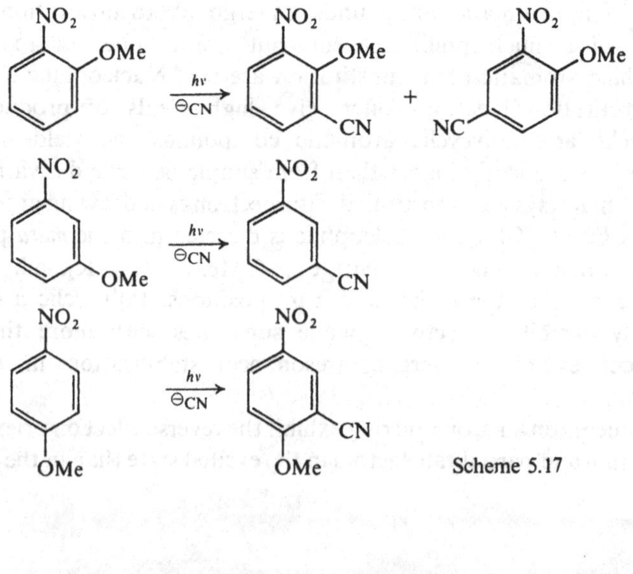

Scheme 5.17

(41)

(42)

'extended' *meta*-activation is observed. These examples serve to demonstrate that the chemistry of the photoexcited molecules contrasts with their ground state chemistry.

The regioselectivity observed in these reactions can be rationalised by the charge distribution of the electronically excited arene, the relative energy of the HOMOs and LUMOs of substrate and nucleophile and the energy surface between reactants and product.[20,21] The reactions generally proceed from the triplet excited arene and are classified as $S_N 2^3 Ar^*$, *viz.* bimolecular nucleophilic substitution involving triplet excited aromatic. Such a reaction sequence is indicated schematically below. The energy gap between the ground and excited surfaces is smallest for the geometry that

$$^{S_0}ArX \xrightarrow{hv} {}^{S_1}ArX \xrightarrow{ISC} {}^{T_1}ArX \xrightarrow{Nu^\ominus} {}^{S_0}[ArXNu]^\ominus \xrightarrow{-X^\ominus} ArNu + X^\ominus$$

leads to the *meta*-σ-complex and hence to *meta*-substitution. The alkaline photohydrolysis of 3,5-dinitroanisole has been examined by flash photolytic techniques and the $S_N Ar^*$ reaction is complete within 10^{-6} s.

σ-complex

The photocyanation of 2-nitrothiophene is an $S_N 2^3 Ar^*$ process. The analogous reaction with 2-nitrofuran is dependent upon the cyanide concentration but the quantum yield of disappearance of nitrofuran is *independent* of cyanide concentration and reaction with other nucleophiles present is competitive. This reaction occurs by an $S_N A^3 Ar^*$ process (scheme 5.18) involving excitation, ISC to the triplet state, and deactivation with ionisation. Examples of this $S_N 1^3 Ar^*$ reaction type are less common than their bimolecular counterparts.

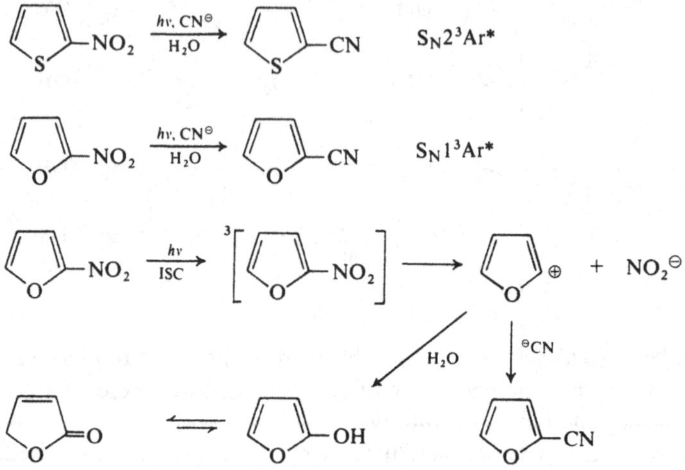

Scheme 5.18

The photocyanation of anisole and its *para*-chloro analogue illustrate the *ortho/para* directing effect of electron-donating substituents. A similar directive influence is observed for 1-methoxynaphthalene which yields the

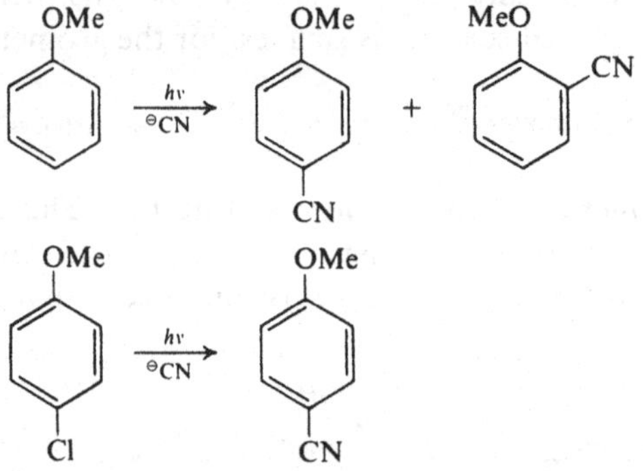

4-cyano-substituted product, and 2-methyoxynaphthalene where the nucleophile is directed into the 1-position. The decisive step of the reactions involves ionisation of triplet excited arene to give a radical cation (scheme 5.19) which reacts with the nucleophile. Such radical cations have been detected spectroscopically from irradiating anisole and the veratroles. Moreover the direction of substitution is consistent with charge-density calculations for the radical cations. Reactions involving the replacement of hydrogen do not take place in the absence of an oxidising agent which suggests that the expulsion of a hydride ion from the intermediate radical (scheme 5.19) is not favoured in the absence of an oxidant. Reactions exhibiting *ortho/para* orientation are classified as $S_{R^+N}1^3Ar^*$ processes.[22]

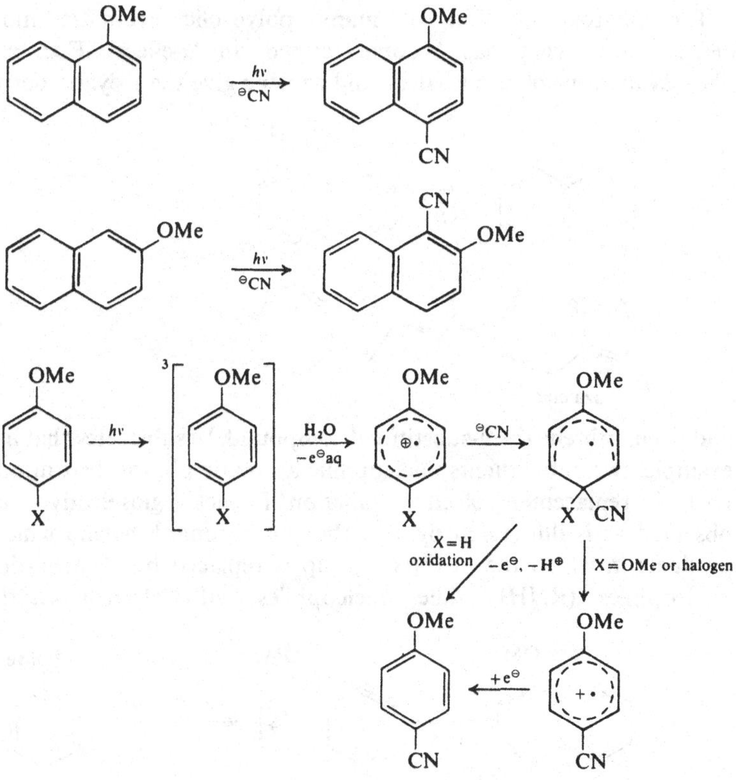

Scheme 5.19

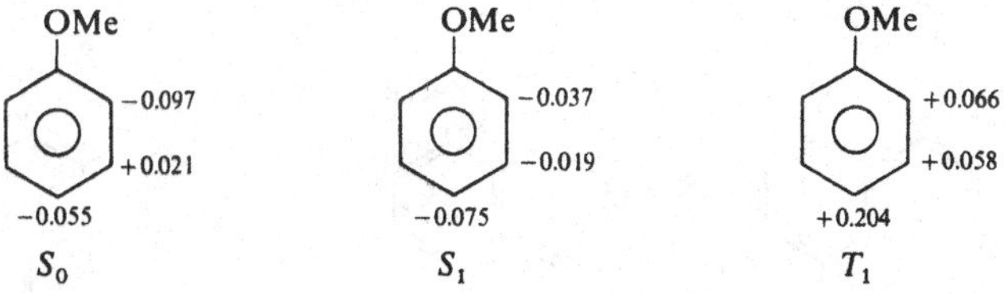

Charge density at *ortho*, *meta* and *para* sites of the S_0, S_1 and T_1 of anisole

The photosubstitution of 2-bromopyridine with the potassium enolates of a variety of ketones in liquid ammonia is a further example of this type of process.

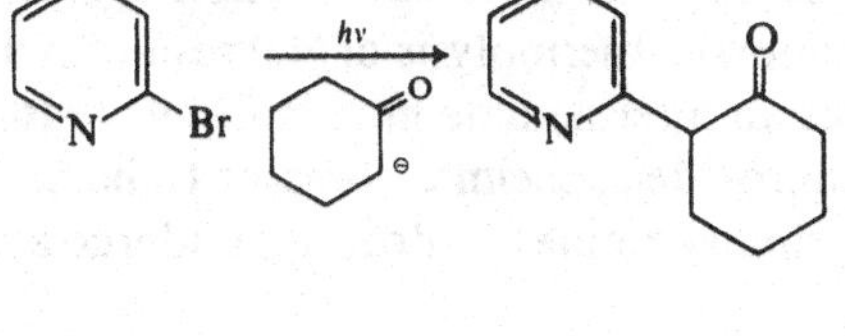

The photosubstitution of many polycyclic aromatic molecules demonstrates what has become termed an 'α-effect'. For example, photocyanations of naphthalene and azulene give the 1-cyano derivatives

and phenanthrene the 9-substituted compound. In substrates that have, for example, two substituents nucleophilic aromatic photosubstitution could result in replacement of either function. In fact regioselectivity is often observed as is illustrated by 1-methoxy-4-nitronaphthalene which gives products in which the methoxy group is replaced by electron-donating nucleophiles (RNH) while nucleophiles with electron-withdrawing

character (CN) replace the nitro group. This is termed merging (resonance) stabilisation[20] since the product formed is that which derives the greatest stabilisation by mesomerism.

Intramolecular aromatic nucleophilic photosubstitution is also known for ω-arylalkylamines. Irradiation or thermolysis of (43) results in a Smiles rearrangement which involves an intermediate intramolecular radical ion pair which rearranges via a spiro-Meisenheimer complex (scheme 5.20) to product. By comparison the primary amine (44) does not undergo a thermal

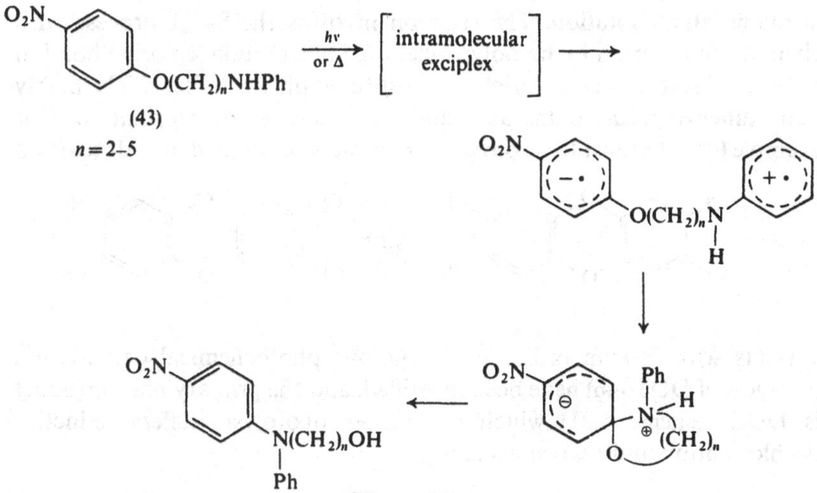

Scheme 5.20

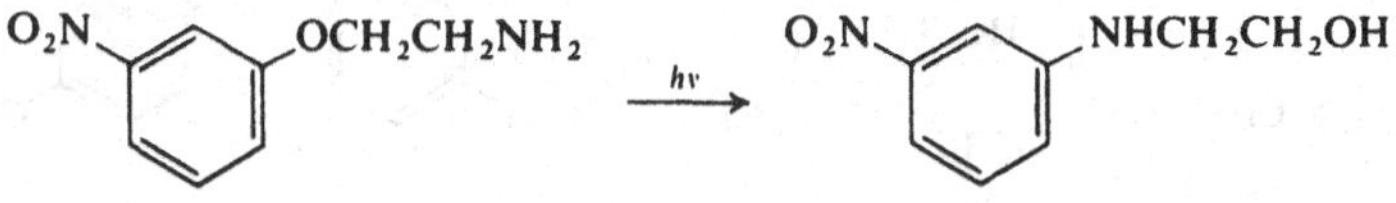

(44)

Smiles rearrangement but the reaction does occur on irradiation. Because electron-transfer interaction between the two nitrogen atoms is no longer favoured due to the absence of the N-phenyl a more direct mechanism involving nucleophilic attack by the amino group on the aromatic ring is postulated.

The photodehalogenation of an aryl chloride to an arene, a reductive dehalogenation, is facilitated by electron donors. For example, *para*-chlorobenzonitrile is reduced to benzonitrile more readily than chlorobenzene is photoreduced to benzene. These reactions are regarded as $S_{R-N}1$ processes[20] and involve unimolecular dissociation of the aryl halide radical anion to an aryl radical and subsequent reaction with nucleophile and electron loss. The photodehalogenation of aryl halides is

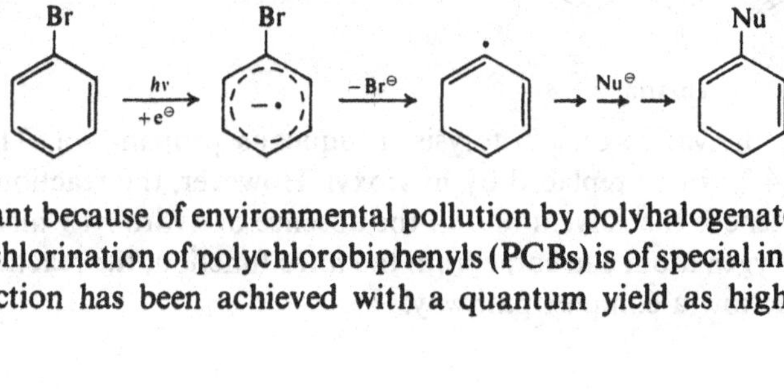

important because of environmental pollution by polyhalogenated arenes. The dechlorination of polychlorobiphenyls (PCBs) is of special interest and the reaction has been achieved with a quantum yield as high as 40 in alcoholic alkali solution. The reaction involves the $S_{R-N}1$ process with chain dechlorination by homolytic cleavage of a carbon–chlorine bond in the excited state to give a variety of substituted phenyl radicals. The highly toxic dibenzo-*para*-dioxins also undergo reductive dechlorination. For example (45), the most toxic dioxin, is mono-dechlorinated in hydrocarbon

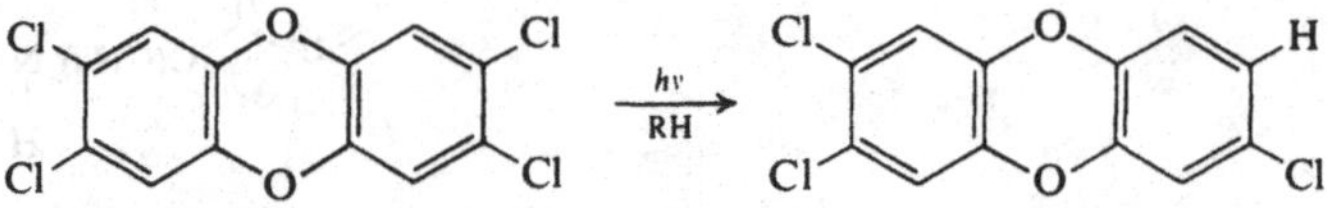

(45)

solvents with 254 nm radiation. Numerous photochemical degradation products of DDT (46) have been identified, and the primary photoproduct is DDE (scheme 5.21) which on further photolysis suffers reductive dechlorination and rearrangement.

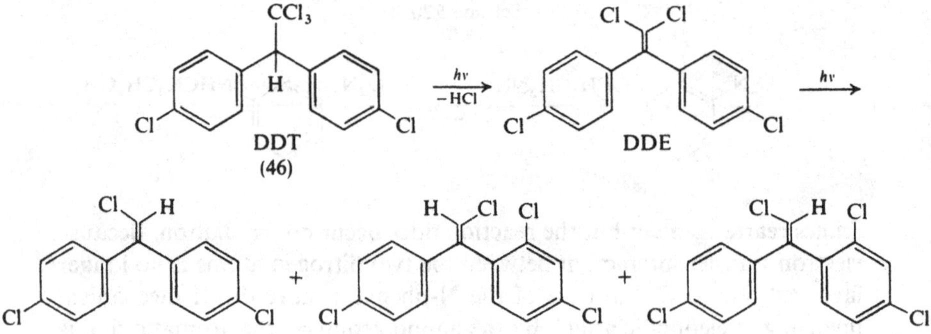

DDT
(46)

DDE

Scheme 5.21

Nucleophilic photosubstitution in aromatic rings of quinones has attracted attention because of its relevance to the light-fastness of anthraquinone dyes (e.g. alizarin). Anthraquinone gives predominantly the

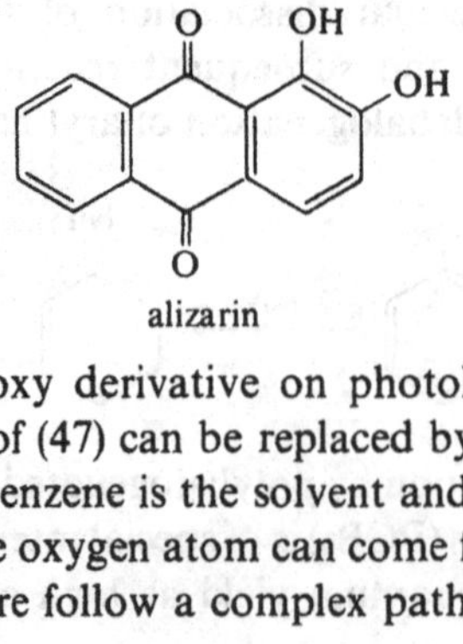

alizarin

2-hydroxy derivative on photolysis in aqueous propan-2-ol. The nitro group of (47) can be replaced by hydroxyl. However, the reaction occurs when benzene is the solvent and in the absence of hydroxyl which shows that the oxygen atom can come from the nitro function; the reaction must therefore follow a complex pathway.

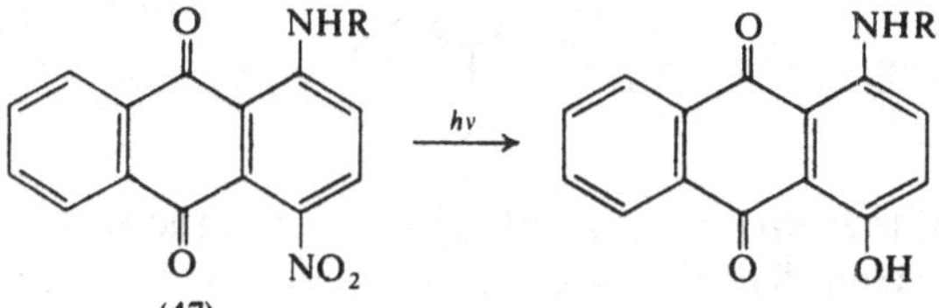

(47)

Some nucleophilic aromatic photosubstitution reactions give the product which would be expected from a ground state reaction, *viz.* electron-withdrawing groups direct *ortho/para*. Such reactions often correspond to substitution reactions which proceed only with difficulty

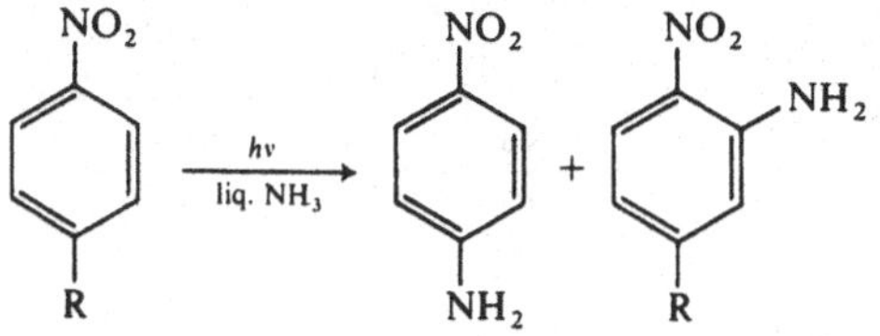

R = H or Cl

under thermal conditions. Thus nitrobenzene and 4-chloronitrobenzene are photoaminated in liquid ammonia with enhanced reactivity at the *ortho*- and *para*-positions, and 3-chloronitrobenzene undergoes photoamination in the 4-position as dictated by the ground state effect of the electron-withdrawing nitro substituent. The reaction mechanism

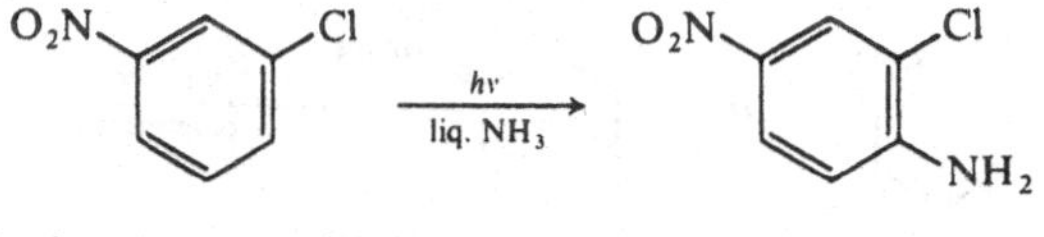

probably involves rapid internal conversion (IC) from the first excited singlet state into a high vibrational level of the ground state and in this way substitution would parallel ground state chemistry.

Aromatic substitution in the ground state is dominated by electrophilic processes which have been intensively investigated and find wide application. By comparison examples of electrophilic photosubstitution reactions are rare. However photodeuteration is important and electron-donating substituents such as methyl or methoxyl direct the electrophile (D⁺) preferentially to the *meta*-position while electron-withdrawing

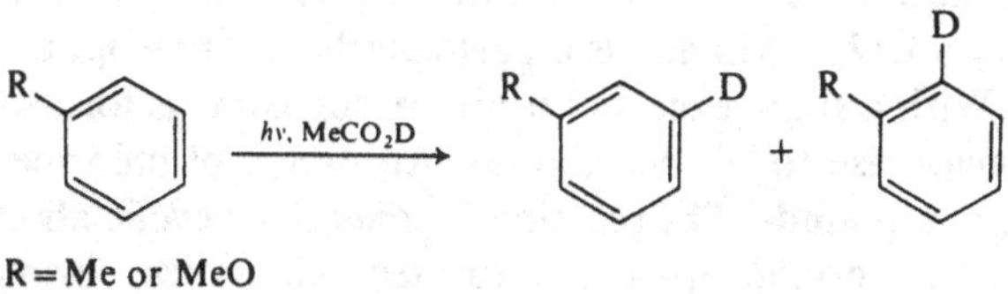

R = Me or MeO

substituents such as the nitro group direct substitution to the *para*-position. The calculated electron densities at the ring carbons in the first excited singlet state show that electron-donating substituents will favour *meta*-substitution if the reaction is fast and controlled by the electron density. The distribution of electron density in the excited singlet state results in a significant change in pK_a while the pK_a of the triplet state is closer to that of the ground state molecule. This is shown for 2-naphthol and 2-naphthoic acid.

		pK_a	
	S_0	S_1	T_1
(OH)	9.5	3.0	8.1
(CO_2H)	4.2	6.6	4.0

Aliphatic and alicyclic molecules are photosubstituted with nitric oxide and the reaction is of industrial importance[23] for the synthesis of ε-caprolactam which is an intermediate in the manufacture of nylon 6.

ε-caprolactam

nylon 6

Nitrosyl chloride and cyclohexane react photochemically to produce oxime, which, on subsequent Beckmann rearrangement, gives ε-caprolactam. The nitrosyl chloride is generated from nitric oxide and chlorine immediately before the gases enter the reaction vessel. A thallium doped mercury arc (λ_{max} 535 nm) is a particularly effective light source for this reaction. With a slight excess of nitric oxide, oxime is formed, while a large excess gives rise to nitroso dimers. An excess of chlorine leads to chloronitroso compounds. The reaction is general for cyclic alkanes (with the exception of cyclopropane), saturated aliphatics, bridged ring hydrocarbons and phenyl-substituted alkanes. The attack of nitric oxide occurs almost exclusively at a methylene site and this is illustrated by the following examples. For alkylbenzenes preference is shown for attack at the

benzylic position. The reaction is effected by light of wavelengths between 600 nm and 200 nm, but the reaction is cleanest when the shorter wavelength radiation (< 500 nm) is removed by a suitable filter. At long wavelengths, a cage four-centre transition structure of alkane and excited nitrosyl chloride is involved, but with shorter wavelength (higher energy) radiation, the reaction exhibits free radical characteristics and chloronitroso compounds are formed as by-products. Analogous photosubstitution reactions can be used to introduce cyano, chlorosulphenyl (–SCl) and dialkoxy phosphinyl ($PO(OR)_2$) (scheme 5.22).

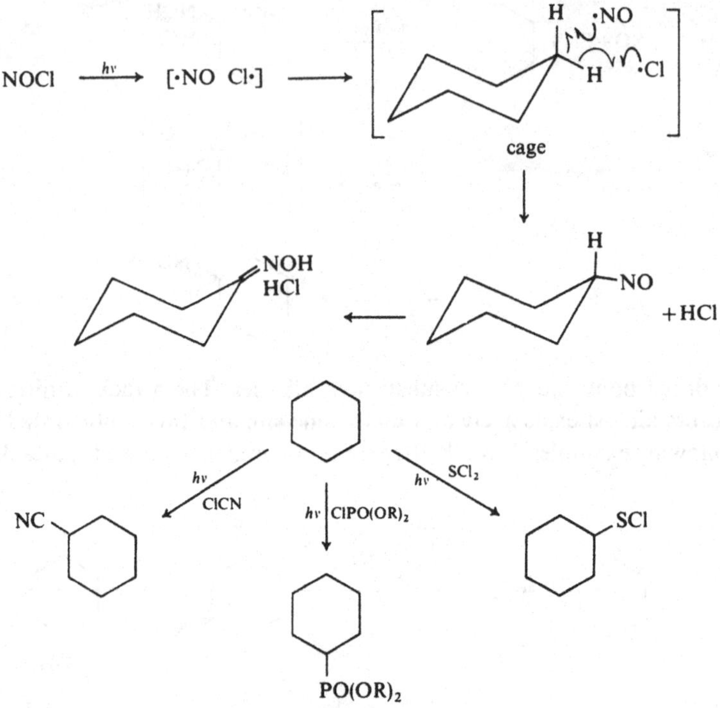

$$NOCl \xrightarrow{h\nu} [\cdot NO \ Cl\cdot] \longrightarrow$$

cage

NOH
HCl

H
NO
+HCl

hν
ClCN

hν ClPO(OR)₂

hν SCl₂

NC

SCl

PO(OR)₂

Scheme 5.22

5.5   Molecular rearrangements involving elimination and substitution[24-26]

On photolysis, organic nitrites can rearrange to oximes or nitroso dimers. Irradiation at wavelengths greater than 330 nm effects homolytic cleavage of the O–NO bond of the nitrite to nitric oxide and alkoxy radical,

$$R\!-\!O\!-\!NO \xrightarrow{h\nu} R\!-\!O\cdot + NO$$

the latter stabilising itself by one of the six routes (scheme 5.23 $(a)–(f)$) available to it. In the presence of a γ-hydrogen atom, intramolecular hydrogen abstraction (route (b)) predominates. The nitric oxide generated in the photolytic step can recombine with the new alkyl radical to produce a nitroso alcohol which can dimerise or, in the presence of protic solvents, form an oxime (scheme 5.24). The reaction involving nitric oxide migration is known as the *Barton reaction* and provides a useful procedure for effecting substitution δ to an oxygen function.[24-26] Provided there is a suitably positioned hydrogen atom the reaction is general for primary and secondary aliphatic nitrites. Thermally generated alkoxy radicals do not

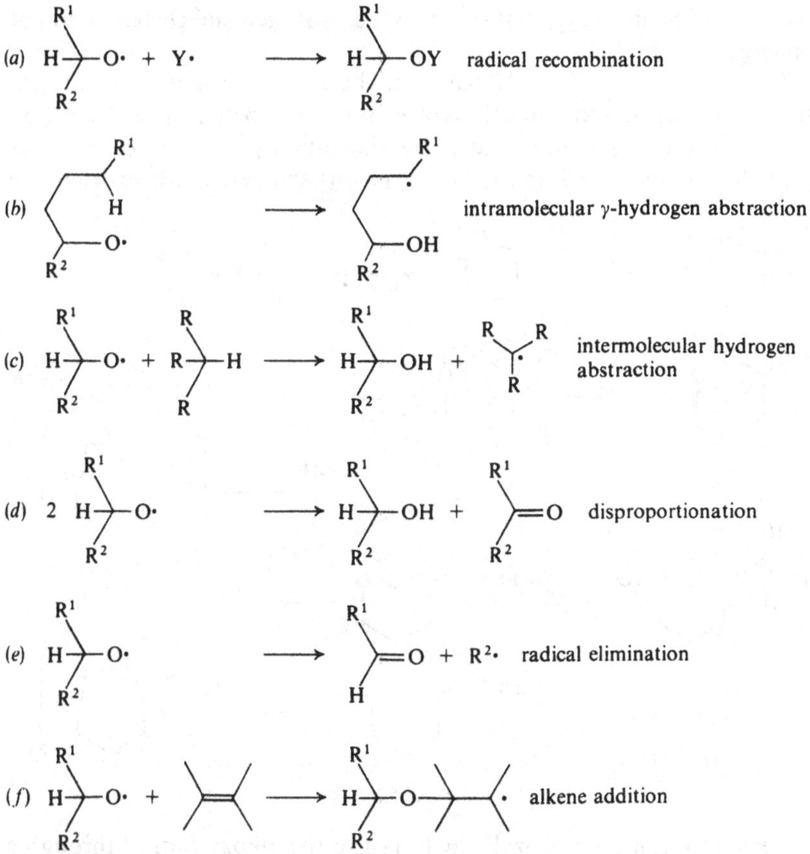

(a) $\overset{R^1}{\underset{R^2}{H{-}\!\!\!\diagup\!\!\!\!\diagdown}}\!\!-O{\cdot} \ + \ Y{\cdot} \quad\longrightarrow\quad \overset{R^1}{\underset{R^2}{H{-}\!\!\!\diagup\!\!\!\!\diagdown}}\!\!-OY$ radical recombination

(b) intramolecular γ-hydrogen abstraction

(c) $\overset{R^1}{\underset{R^2}{H{-}\!\!\!\diagup\!\!\!\!\diagdown}}\!\!-O{\cdot} \ + \ \overset{R}{\underset{R}{R{-}\!\!\!\diagup\!\!\!\!\diagdown}}\!\!-H \ \longrightarrow \ \overset{R^1}{\underset{R^2}{H{-}\!\!\!\diagup\!\!\!\!\diagdown}}\!\!-OH \ + \ $ intermolecular hydrogen abstraction

(d) $2\ \overset{R^1}{\underset{R^2}{H{-}\!\!\!\diagup\!\!\!\!\diagdown}}\!\!-O{\cdot} \quad\longrightarrow\quad \overset{R^1}{\underset{R^2}{H{-}\!\!\!\diagup\!\!\!\!\diagdown}}\!\!-OH \ + \ \overset{R^1}{\underset{R^2}{\diagup\!\!\!\!\diagdown}}\!\!\!=O$ disproportionation

(e) $\overset{R^1}{\underset{R^2}{H{-}\!\!\!\diagup\!\!\!\!\diagdown}}\!\!-O{\cdot} \quad\longrightarrow\quad \overset{R^1}{\underset{H}{\diagup\!\!\!\!\diagdown}}\!\!\!=O \ + \ R^2{\cdot}$ radical elimination

(f) $\overset{R^1}{\underset{R^2}{H{-}\!\!\!\diagup\!\!\!\!\diagdown}}\!\!-O{\cdot} \ + \ \searrow\!\!=\!\!\swarrow \ \longrightarrow \ \overset{R^1}{\underset{R^2}{H{-}\!\!\!\diagup\!\!\!\!\diagdown}}\!\!-O\!\!-\!\!\diagdown\!\!\!\cdot$ alkene addition

Scheme 5.23

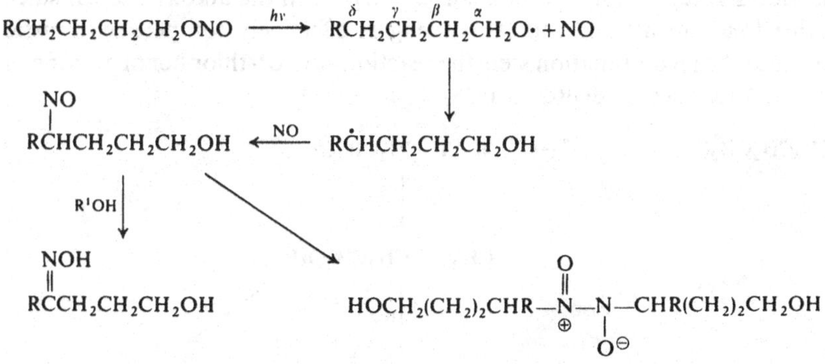

$$RCH_2CH_2CH_2CH_2ONO \xrightarrow{\ h\nu\ } R\overset{\delta}{C}H_2\overset{\gamma}{C}H_2\overset{\beta}{C}H_2\overset{\alpha}{C}H_2O{\cdot} + NO$$

$$\downarrow$$

$$\underset{|\,NO}{RCHCH_2CH_2CH_2OH} \xleftarrow{\ NO\ } R\dot{C}HCH_2CH_2CH_2OH$$

$$R^1OH\downarrow \qquad\qquad \searrow$$

$$\underset{\overset{\|}{\text{NOH}}}{RCCH_2CH_2CH_2OH} \qquad\qquad HOCH_2(CH_2)_2CHR\!-\!\overset{\overset{O}{\|}}{\underset{\underset{O^{\ominus}}{|}}{N}}\!-\!N\!-\!CHR(CH_2)_2CH_2OH$$

$$\overset{\oplus}{}$$

Scheme 5.24 nitroso dimer

possess sufficient energy to abstract such a hydrogen and therefore do not undergo this reaction.

Like the Norrish Type II reaction, the intramolecular 1,5-hydrogen transfer to the alkoxy radical involves a six-membered cyclic transition state structure, as demonstrated by the photolysis of a series of ω-phenylalkyl nitrites (scheme 5.25). The propyl and pentyl nitrites might be

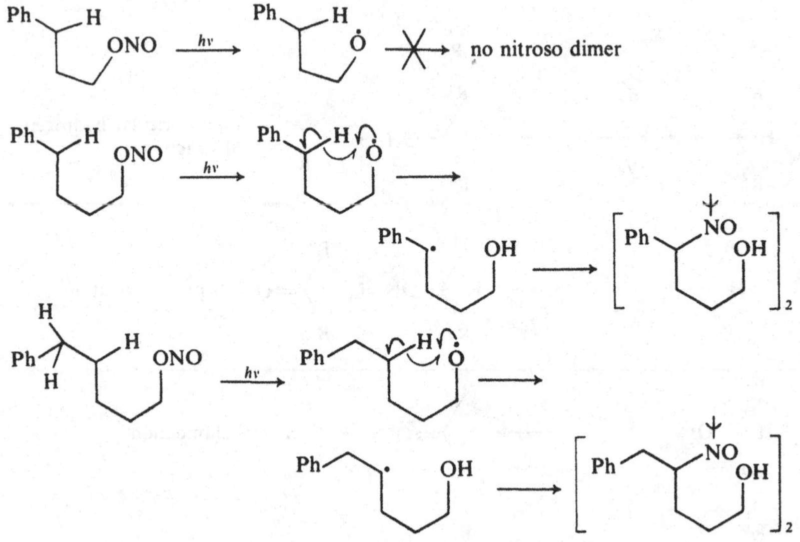

Scheme 5.25

expected to transfer a benzylic hydrogen to the alkoxy radical through a five- and a seven-membered cyclic transition state structure but this is not observed. 1,5-Hydrogen transfer is observed for the pentyl nitrites and similarly for butyl nitrites. In general the nitric oxide produced in the homolysis step is not held in a solvent cage with the alkoxy radical since other free radicals and radical scavengers effectively compete with nitric oxide in the recombination step; the reaction with d^1-thiophenol provides a method for specific deuteration.

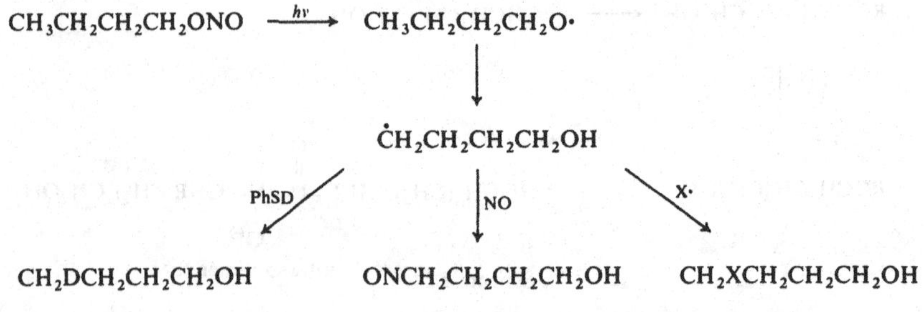

For tertiary nitrites, and primary and secondary nitrites with no available hydrogen the alkoxy radical is generally stabilised by inter-molecular hydrogen abstraction (path (c), scheme 5.23), disproportionation (path (d), scheme 5.23), or radical elimination (path (e), scheme 5.23). The product of photolysis of an alicyclic nitrite is dependent upon the size of the ring (scheme 5.26). Ring cleavage to a nitroso-aldehyde results whenever a

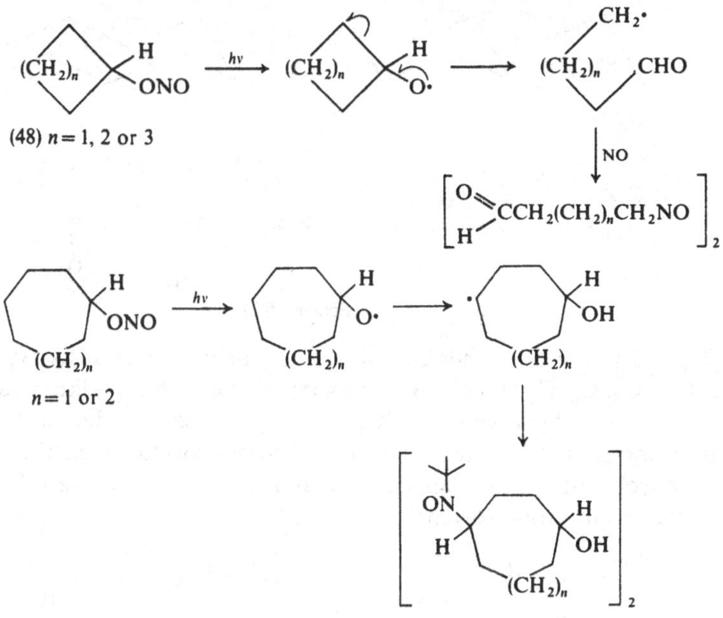

$(CH_2)_n$ H ONO $\xrightarrow{h\nu}$ $(CH_2)_n$ H O· $\longrightarrow$ $(CH_2)_n$ CH$_2$· CHO

(48) $n = 1$, 2 or 3

$\downarrow$ NO

$$\left[\begin{matrix} O \\ \diagdown \\ H \end{matrix} CCH_2(CH_2)_nCH_2NO \right]_2$$

$(CH_2)_n$ H ONO $\xrightarrow{h\nu}$ $(CH_2)_n$ H O· $\longrightarrow$ $(CH_2)_n$ H OH

$n = 1$ or 2

$$\left[\begin{matrix} ON \\ \\ H \end{matrix} \quad \begin{matrix} H \\ OH \\ (CH_2)_n \end{matrix} \right]_2$$

Scheme 5.26

cyclic six-membered transition structure for hydrogen transfer cannot be attained (48, $n \leqslant 3$), but 1,5-hydrogen transfer occurs in seven- and larger-membered ring systems.

The stereochemical requirements for the cyclic transition structure for hydrogen transfer are demonstrated by the substituted cyclohexyl nitrites shown in scheme 5.27. While *cis*- and *trans*-2-ethylcyclohexyl nitrite undergo 1,5-hydrogen transfer reactions, *cis*-3-methylcyclohexyl nitrite does not. The alkoxy radical formed on photolysis of this latter compound is too far removed for 1,5-hydrogen transfer and the barrier to ring inversion to the diaxial conformer is such that it is not attained (scheme 5.27).

The Barton reaction has found extensive use in functionalising the angular methyl groups of steroids[24–26] and these reactions further illustrate the stereochemical requirements for 1,5-hydrogen transfer. The C_{18} angular methyl group (at C_{13}) can be attacked by an alkoxy radical at

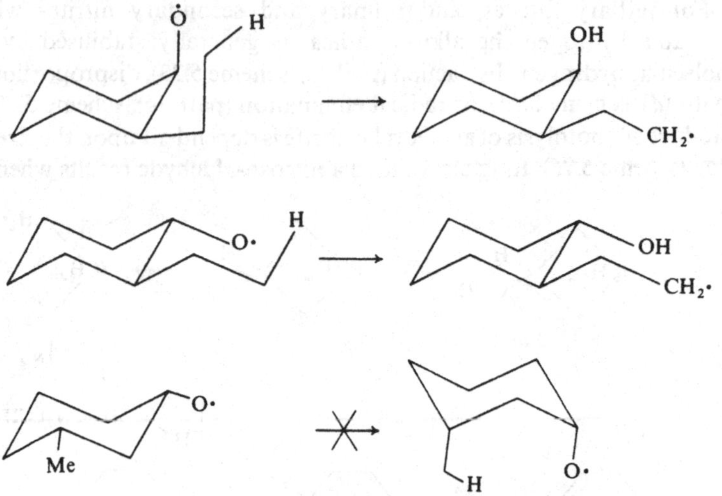

Scheme 5.27

C_{20}, C_{15}, C_{11} and C_8 while C_{19} (at C_{10}) may react with axial alkoxy radicals at C_{11}, C_8, C_6, C_4 and C_2. By comparison an equatorial alkoxy radical at C_{11} abstracts hydrogen from C_1. These processes are demonstrated for aldoxime formation from 3β-acetoxy-5α-pregnan-20β-ol nitrite (49), 3β-acetoxycholestan-6β-ol nitrite (50) and 11α-hydroxycholest-4-en-3-one nitrite (51) in protic solvents.

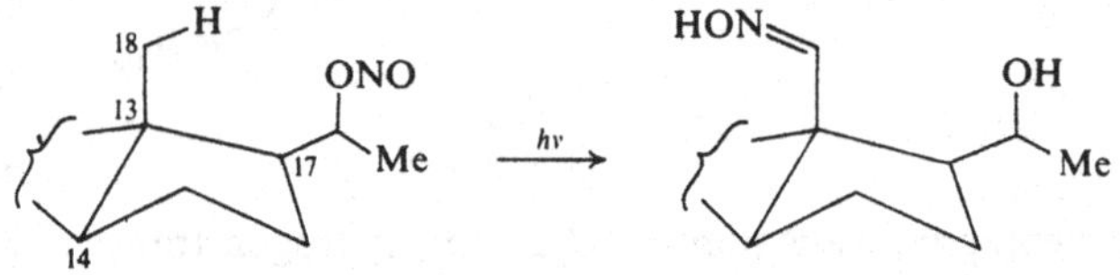

(49)

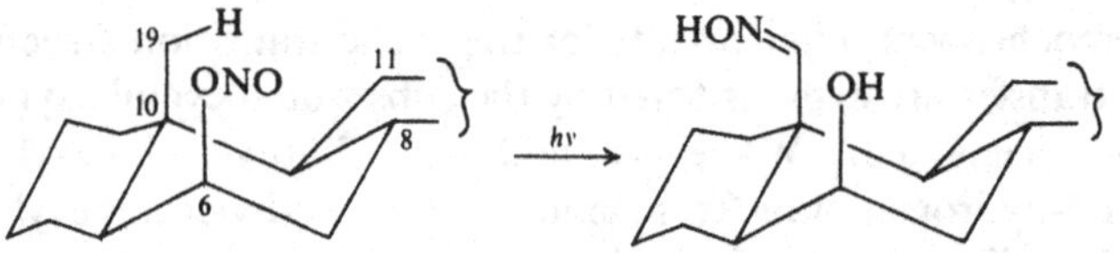

(50)

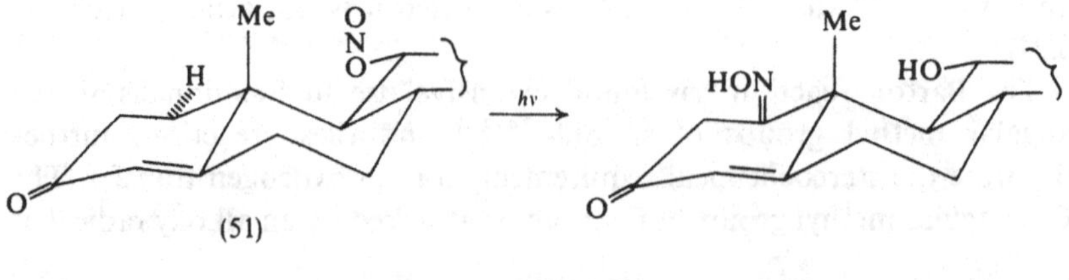

(51)

Since radical recombination is not a cage process in the Barton reaction, the alkyl radical produced can follow alternative lower energy stabilisation paths. The corticosterone derivative (52), the 11β-isomer of (51) (above), gives the 4-ketoxime (54) rather than the 19-aldoxime (53).

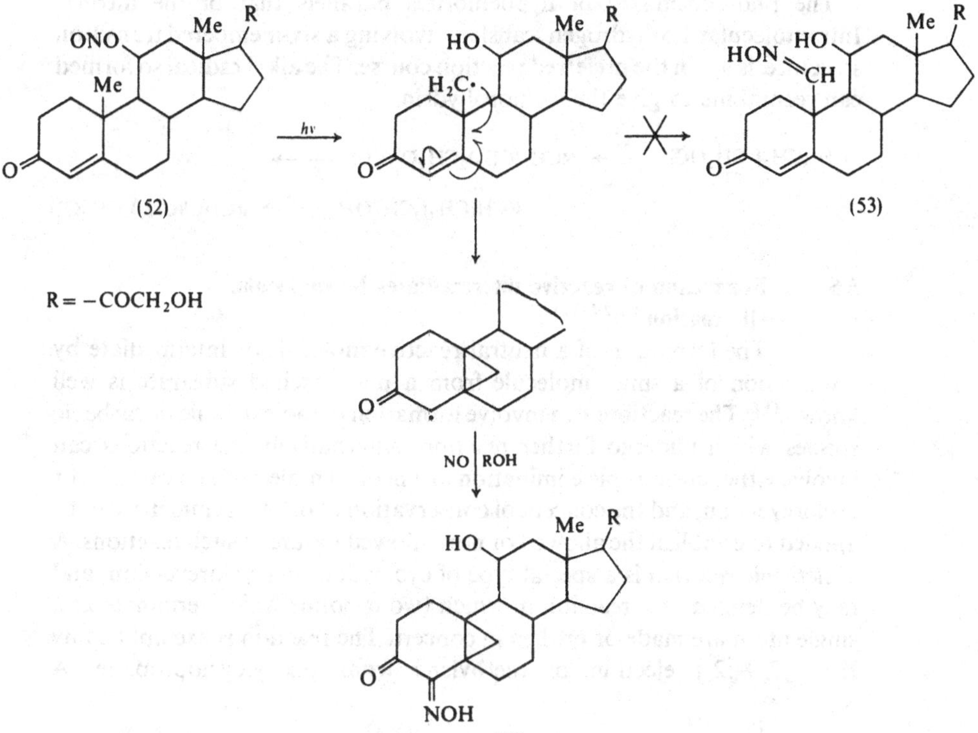

(52)

R = —COCH₂OH

(53)

(54)

In the presence of copper(II) acetate, radical recombination is prevented and oxidation to an enol ensues. Thus hexan-2-ol nitrite is converted into a mixture of unsaturated alcohols on photolysis. A further variation to the reaction can occur with γδ-unsaturated nitrites whereupon radical ring closure produces a tetrahydrofuranyl derivative.

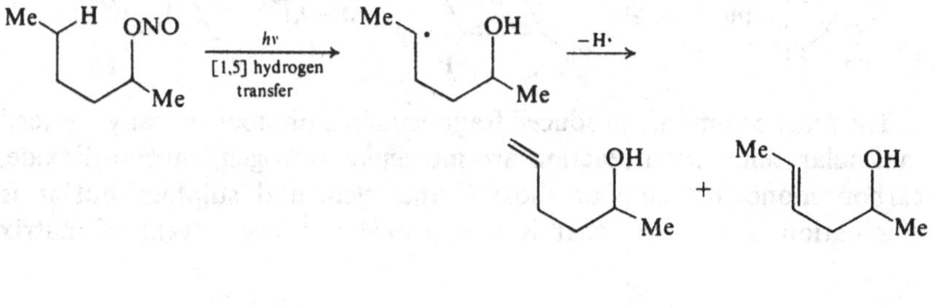

The photochemistry of hypochlorites parallels that of the nitrites. Intramolecular 1,5-hydrogen transfer, involving a six-membered transition structure, is again the preferred reaction course. The alkyl radical so formed can recombine to give the γ-chlorohydrin.

$$RCH_2(CH_2)_2CH_2OCl \xrightarrow{h\nu} RCH_2(CH_2)_2CH_2O\cdot + Cl\cdot \longrightarrow$$

$$R\overset{\cdot}{C}H(CH_2)_2CH_2OH \xrightarrow{Cl\cdot} RCHCl(CH_2)_2CH_2OH$$

5.6   Formation of reactive intermediates by molecular elimination[14,27-32]

The formation of a neutral reactive molecule or intermediate by the ejection of a small molecule from a photoexcited substrate is well known.[14] The reactions can involve formation of radical, ionic or carbenic species which undergo further reaction. Alternatively the reactions can involve either cheletropic elimination of a neutral molecule or a carbene, or cycloreversion, and the concept of conservation of orbital symmetry can be applied to establish the allowed or non-allowed nature of such reactions. A *cheletropic reaction* is a special type of cycloaddition (cycloreversion) and may be defined as a reaction in which two σ-bonds which terminate at a single atom are made or broken in concert. The reaction is exemplified by the $[_\sigma 2_s + _\sigma 2_s]$ ejection of methylene from phenylcyclopropane. A

cycloreversion is the reverse process to cycloaddition (chapter 4) and the $[_\sigma 2_s + _\sigma 2_s]$ cleavage of tetraphenylquadricyclene illustrates such reactions.

The most commonly produced fragments in a photochemically induced molecular elimination reaction are molecular nitrogen, carbon dioxide, carbon monoxide, sulphur dioxide and elemental sulphur, but it is elimination of nitrogen that is best known.[14] The advent of matrix isolation procedures, and the availability of low temperature spectroscopic techniques, has resulted in an upsurge of interest in these reactions and a large number of reactive molecules have now been prepared and characterised by these techniques.

The photochemical decomposition of azoalkanes provides a general and convenient route to alkyl radicals which complements the thermal counterpart as a route to alkyl radicals. While decomposition of the energy-rich excited state occurs for simple azoalkanes, the thermal process requires the presence of groups within the molecule that can stabilise the free radical

$$\text{Me}_2\text{CH} \diagdown \text{N}{=}\text{N} \diagdown \text{CHMe}_2 \xrightarrow{hv} 2\text{Me}_2\dot{\text{C}}\text{H} + \text{N}_2$$

$$\underset{\overset{|}{\text{CN}}}{\overset{\text{CN}}{\overset{|}{\text{Me}_2\text{C}}}} \diagdown \text{N}{=}\text{N} \diagdown \underset{\overset{|}{\text{CN}}}{\text{CMe}_2} \xrightarrow[80\ \text{C}]{\Delta} 2\text{Me}_2\dot{\text{C}}\text{CN} + \text{N}_2$$

in order for the reaction to proceed at moderate temperatures. The loss of nitrogen from azoalkanes also occurs for bicyclic species to give a cyclic diradical. In general, these diradicals afford cycloalkanes or their derivatives as shown for the formation of bicyclo[2,2,0]hexane. The unsaturated analogue is more reactive losing nitrogen in a thermally-allowed $[_\sigma2_s + _\sigma2_s + _\pi2_s]$ cycloreversion to give cyclohexa-1,3-diene.

Photodeazetation of pyrazolines provides a general synthesis of cyclopropanes as illustrated for the highly strained tetracycle (55)[27] and the propellane (56).[28] With methylenepyrazolines and their derivatives, trimethylenemethane diradicals are produced on photolysis and products of rearrangement are observed (scheme 5.28). In the bicyclic framework (scheme 5.29) the diradical intermediate is a singlet species and has been trapped with acrylonitrile.[29]

The smallest cyclic azo compound, diazirine, suffers nitrogen loss on photolysis to give a carbene. The $[_\sigma2_s + _\sigma2_s]$ reaction to give carbenes is general for a variety of substituted diazirines as shown in scheme 5.30.[30] The generation of carbenes by photodeazetation more commonly employs readily accessible diazo compounds as substrates. Methylene itself is

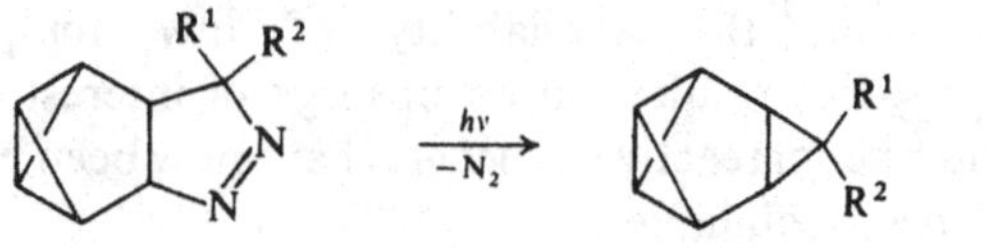

(55)

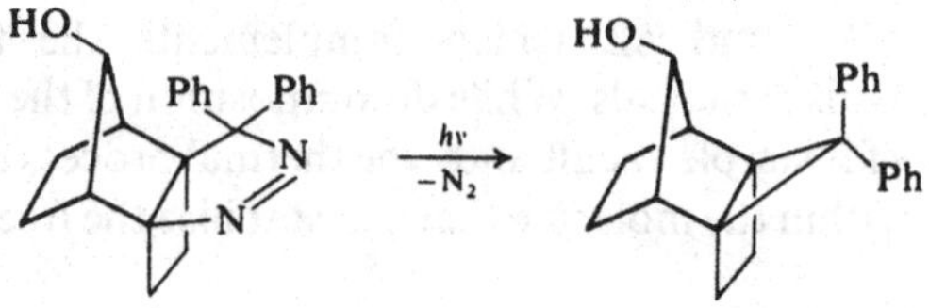

(56)

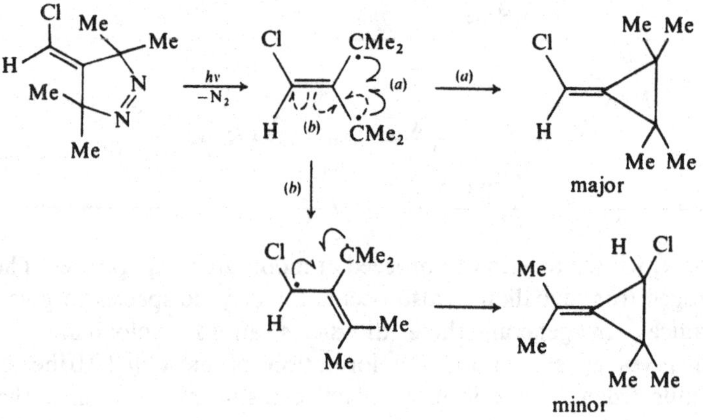

Scheme 5.28

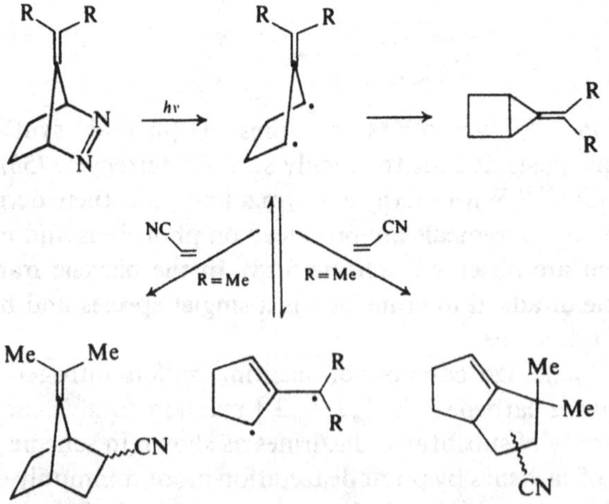

Scheme 5.29

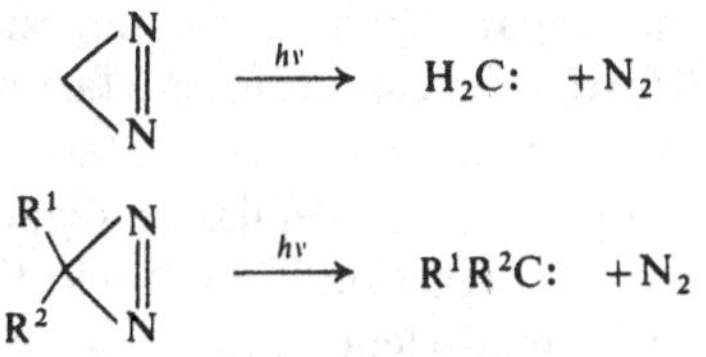

$R^1 = Cl; R^2 = CF_3$ or cyclo-C_3H_5
Scheme 5.30

obtained both by thermolysis and photolysis of diazomethane. When generated by the direct photolysis of diazomethane in isopropanol,

$$CH_2N_2 \xrightarrow[\text{or } h\nu]{\Delta} :CH_2 + N_2$$

methylene is inserted into all possible bonds of the solvent, but in more inert solvents methylene will stereospecifically add to an alkene. The triplet

$$CH_2N_2 \xrightarrow[\text{Me}_2\text{CHOH}]{h\nu} Me_3COH + Me_2CHOMe + MeCH(OH)Et$$

methylene, produced by sensitised photolysis of diazomethane, reacts separately with *cis*- and *trans*-but-2-ene to give non-identical mixtures of

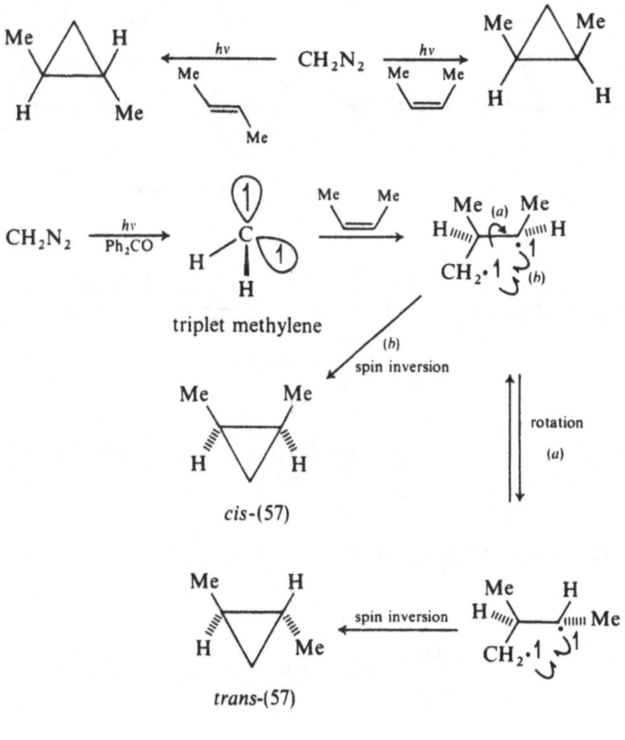

triplet methylene

cis-(57)

trans-(57)

cis- and *trans*-1,2-dimethylcyclopropane (57). This suggests that spin inversion, required before bond formation can occur, is a fast process since carbon–carbon bond formation competes with bond rotation.

Almost all aliphatic diazo compounds are similar to diazomethane in their behaviour. The carbene insertion reaction is more selective the more stable the carbene. For example, methylene formed by photolysis of diazomethane inserts into all the carbon–hydrogen bonds of 2,3-dimethylbutane with practically no stereoselectivity whereas carbomethoxymethylene, a more stable carbene, inserts preferentially into the tertiary carbon–hydrogen bond. Carbenes produced from diazoketones and diazoesters undergo a Wolff rearrangement in protic media to form carboxylic esters, a reaction which competes for these substrates with hydrogen abstraction and carbene insertion. Thus the cyclic α-diazoketone (58) yields a carboxylic ester on direct photolysis even in the presence of an alkene. Furthermore, the reaction does not involve an oxirene intermediate since substrates labelled at C_1 provide ester with no label in the ring. The

major product

$$PhCOCHN_2 \xrightarrow[-N_2]{h\nu} Ph{-}\overset{O}{\overset{\|}{C}}{-}\overset{\cdot\cdot}{C}H \xrightarrow[\text{rearrangement}]{\text{Wolff}} Ph{-}CH{=}C{=}O$$

$$H\cdot \downarrow \text{from solvent} \qquad\qquad \downarrow ROH$$

$$Ph{-}\overset{O}{\overset{\|}{C}}{-}CH_3 \qquad\qquad Ph{-}CH_2{-}C\overset{\diagup O}{\underset{\diagdown OR}{}}$$

$$R^1O_2CCHN_2 \xrightarrow[-N_2]{h\nu} R^1O{-}\overset{O}{\overset{\|}{C}}{-}\overset{\cdot\cdot}{C}H \xrightarrow[\text{rearrangement}]{\text{Wolff}} R^1O{-}CH{=}C{=}O$$

$$\downarrow R^2OH \qquad\qquad \downarrow R^2OH$$

$$\overset{O}{\underset{R^1O}{\diagdown}}C{-}CH_2{-}OR^2 \qquad\qquad R^1O{-}CH_2{-}C\overset{\diagup O}{\underset{\diagdown OR^2}{}}$$

photo-Wolff rearrangement has recently been used in the synthesis of aza-β-lactams which are potential antibacterial agents.[31] A further interesting

example is shown by the bis-diazo compound (59) which on photolysis (λ > 364 nm) ring contracts to give acenaphthyne which has been isolated in

a matrix at 15 K. Carbenes are also produced by photodecomposition of tosylhydrazones or their sodium salts and such a reaction has been cleverly utilised in the synthesis of the [2,2,2,2]cyclophane (60).

Whereas the pyrazolines provide diradicals upon photolysis, their conjugated homologues, the pyrazoles provide αβ-unsaturated carbenes which cyclise to cyclopropenes. The reactions proceed by way of the ring-opened diazo isomer and provide a general synthesis of cyclopropenes.[32] Thus cyclopropene (61) is available by two routes via a common diazo intermediate.

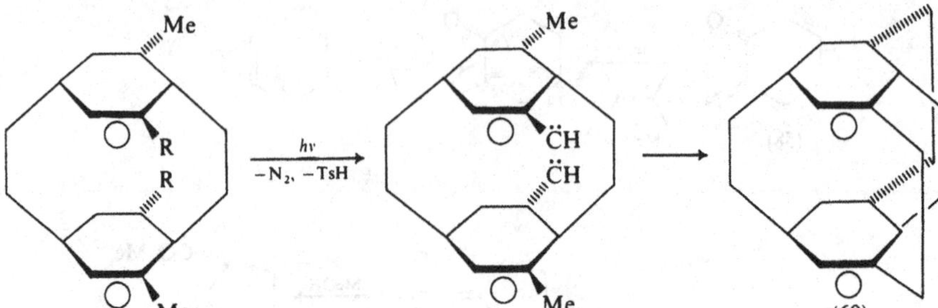

R = −CHNNHTs

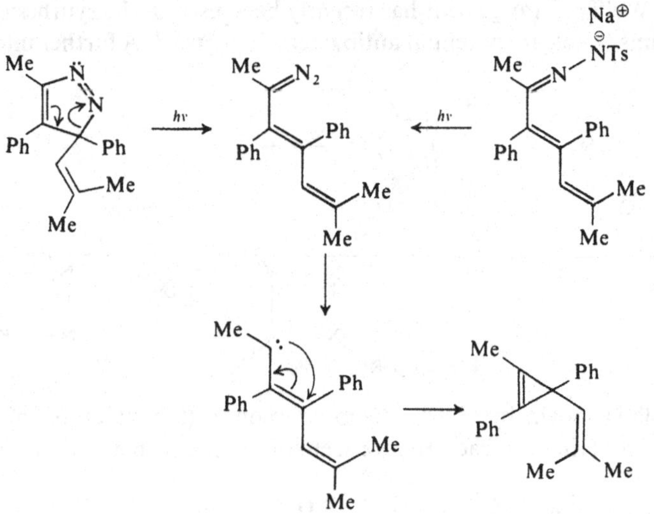

The photochemical loss of nitrogen is not restricted to molecules in which the incipient nitrogen molecule is bonded to carbon. This is shown below for a range of compounds each of which has its analogue in the compound types discussed above.

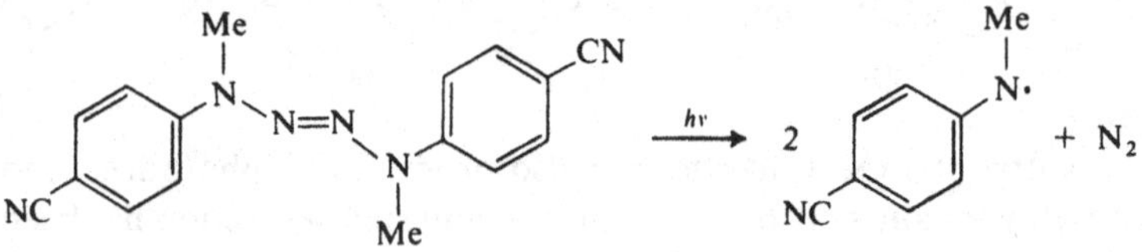

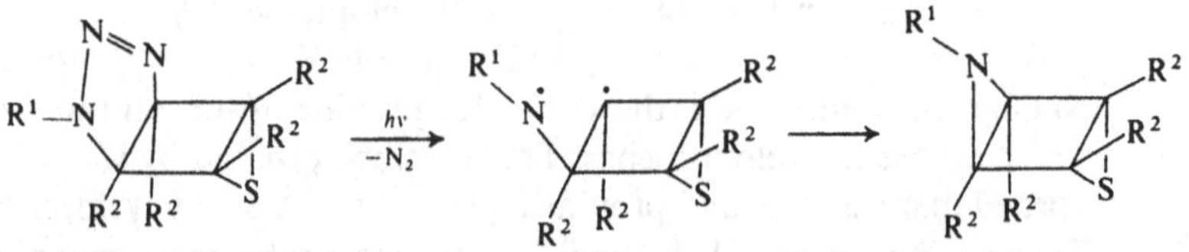

R^1 = Ph, C_6H_{11}, Bu^t
R^2 = CF$_3$

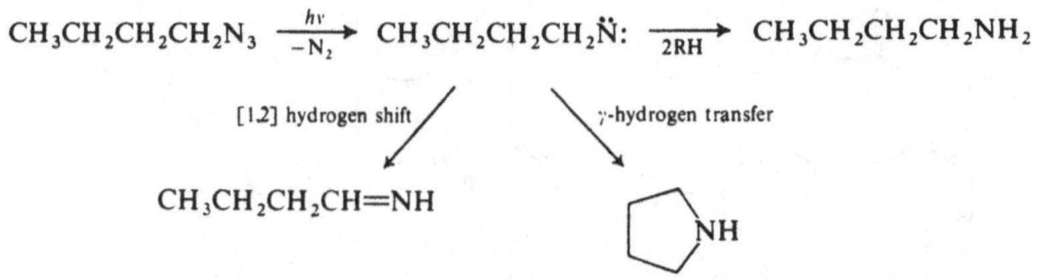

$$CH_3CH_2CH_2CH_2N_3 \xrightarrow[-N_2]{h\nu} CH_3CH_2CH_2CH_2\ddot{N}: \xrightarrow[2RH]{} CH_3CH_2CH_2CH_2NH_2$$

[1,2] hydrogen shift /   \ γ-hydrogen transfer

$$CH_3CH_2CH_2CH=NH$$

NH

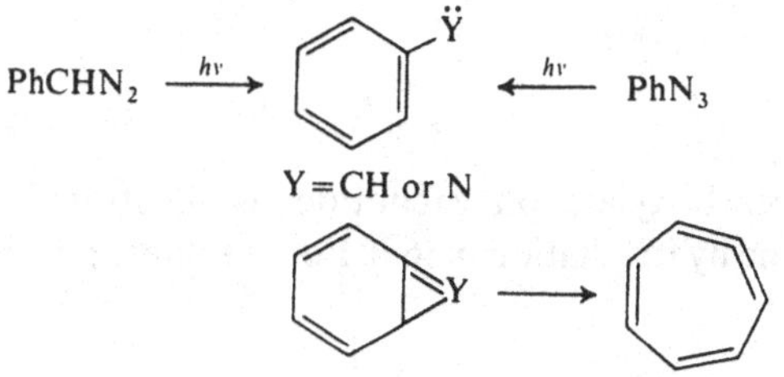

$$PhCHN_2 \xrightarrow{h\nu} \quad \xleftarrow{h\nu} PhN_3$$

Y=CH or N

Y=CH isolable in matrix at 10 K

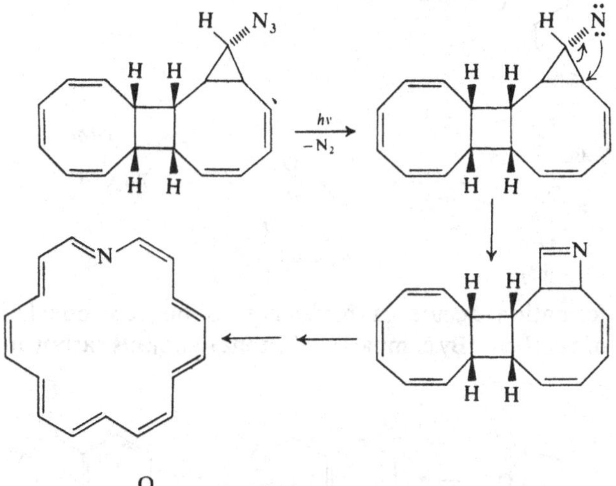

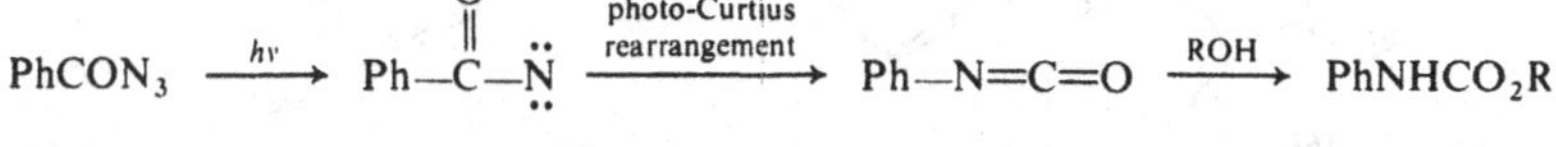

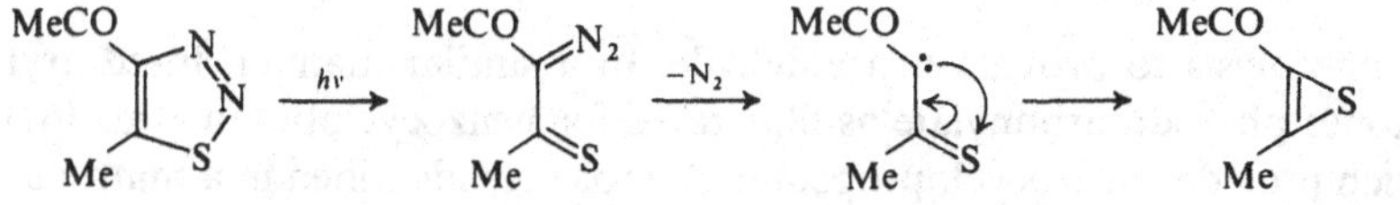

$$PhCON_3 \xrightarrow{h\nu} Ph-\overset{\overset{\displaystyle O}{\|}}{C}-\ddot{\ddot{N}} \xrightarrow[\text{rearrangement}]{\text{photo-Curtius}} Ph-N=C=O \xrightarrow{ROH} PhNHCO_2R$$

The loss of carbon dioxide from the photoexcited state is less common than deazetation. Nonetheless, the reaction is well documented[14] and provides some interesting synthetic applications. For example, [2,2]paracyclophane (63) is available in high yield from the bis-ester (62).

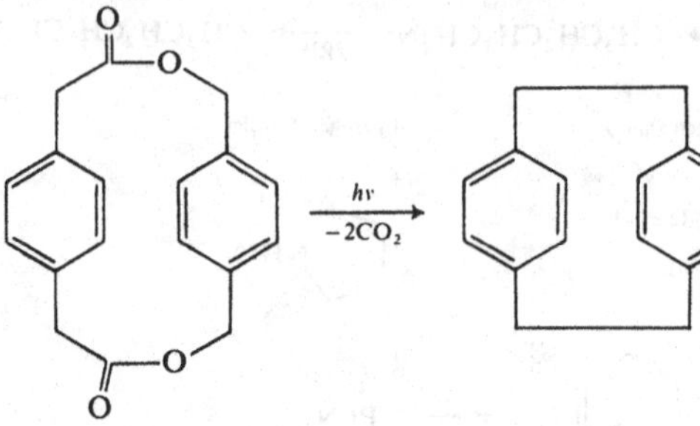

(62)   (63)

Peranhydrides can decarboxylate on excitation as illustrated for the formation of cyclohexene by irradiation of (64). For malonate peroxide (65)

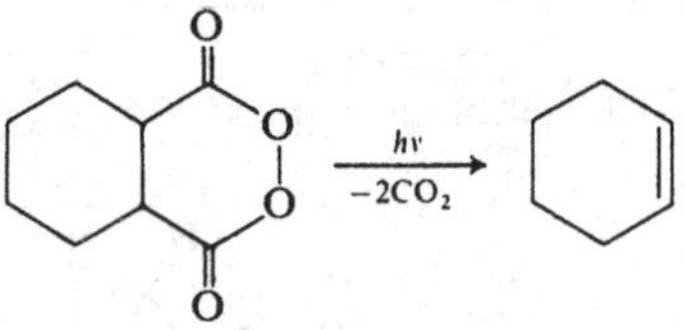

(64)

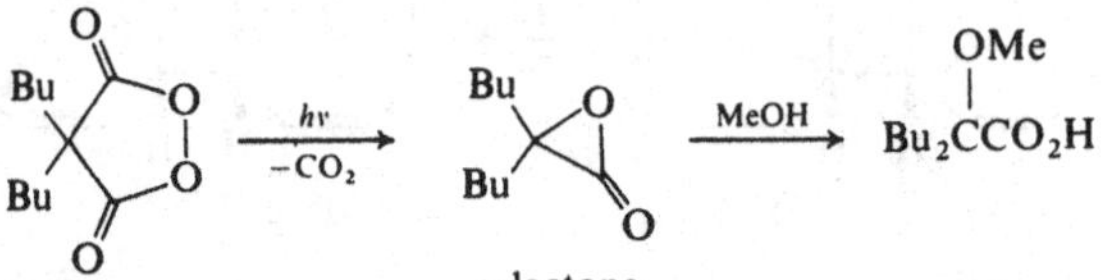

(65)   α-lactone

α-lactone formation occurs in preference to bis-decarboxylation and formation of a carbene. By comparison lactone (66) ejects carbon monoxide

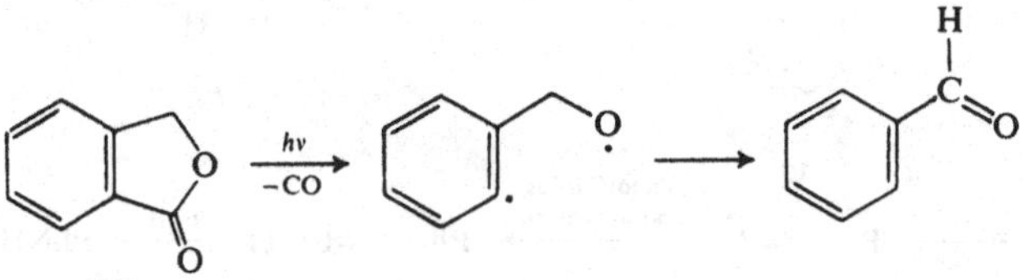

(66)

on photolysis to provide benzaldehyde. In a similar manner, fused aryl ketones photodecarbonylate as illustrated for benzocyclobutandione (67) which provides benzocyclopropenone, a molecule identified in a matrix at 10 K (scheme 5.31). Subsequent photolysis results in loss of a second molecule of carbon monoxide and formation of dehydrobenzene. As illustrated in scheme 5.31, ketenes photochemically decarbonylate with

$$CH_2=C=O \xrightarrow{hv} :CH_2 + CO$$

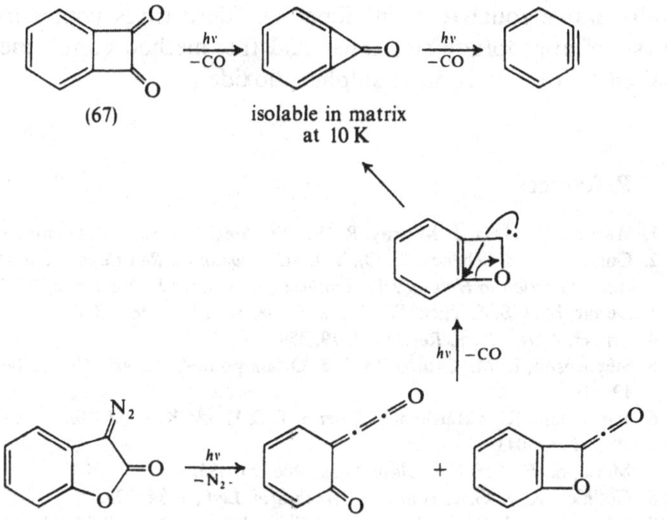

isolable in matrix
at 10 K

Scheme 5.31

concomitant formation of a carbene. In a reaction related to those shown in scheme 5.31, decarbonylation of (68) leads to the heteroquinodimethane

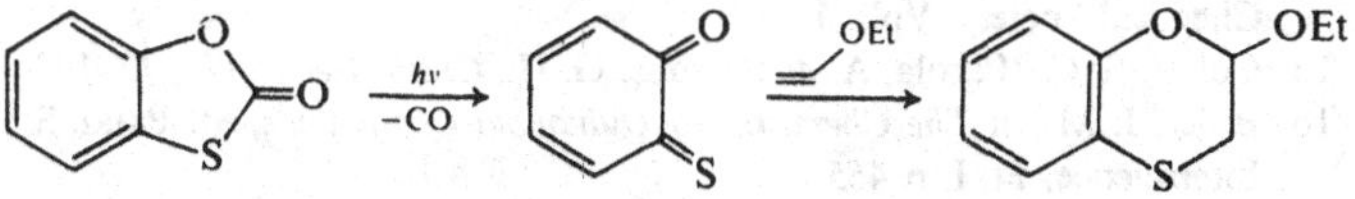

(69) which can be trapped with vinyl ethers in a Diels–Alder cycloaddition.[32] It is interesting to note that decarboxylation is not observed in this case whilst the ejection of carbon disulphide from (70)

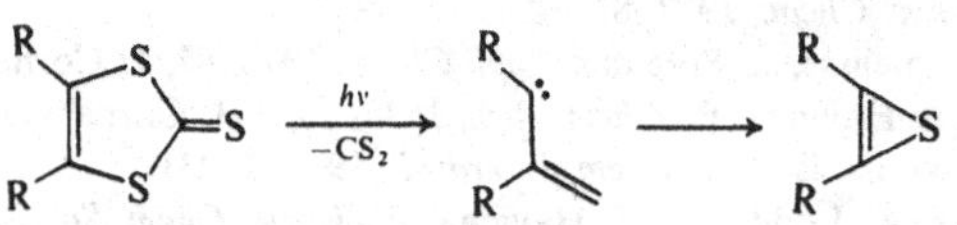

proceeds to give thiirenes which can be isolated by matrix techniques at low temperatures.

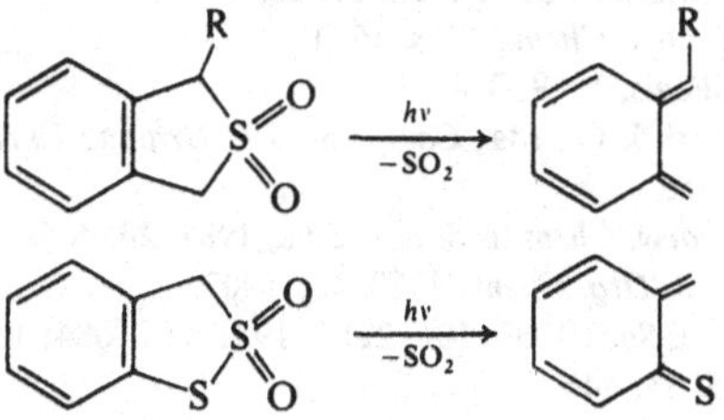

An alternative route to quinodimethane derivatives comes from the photolysis of appropriate sulphones and this method complements the thermal cheletropic ejection of sulphur dioxide.

References

1. Wasserman, H. H. & Murray, R. W., eds., *Singlet Oxygen*, Academic, 1979.
2. Gollnick, K. & Schenck, G. O., in *1,4-Cycloaddition Reactions – The Diels–Alder Reaction in Heterocyclic Synthesis*, ed. Hamer J., Academic, 1967.
3. Dewar, M. J. S. & Theil, W., *J. Am. Chem. Soc.*, 1977, **99**, 2338.
4. Frimer, A. A., *Chem. Rev.*, 1979, **79**, 359.
5. Stephenson, L. M., Grdina, M. J. & Orfanopoulos, M., *Acc. Chem. Res.*, 1980, **13**, 419.
6. Yamaguchi, K., Yabushita, S., Fueno, T. & Houk, K. N., *J. Am. Chem. Soc.*, 1981, **103**, 5043.
7. McCapra, F., *Qtr. Rev. Chem. Soc.*, 1966, **20**, 485.
8. Gollnick, K. & Griesbaum, A., *Tetrahedron Lett.*, 1984, 725.
9. Adam, W. & Baader, W. J., *Angew. Chem. Internat. Edn.*, 1984, **23**, 166.
10. Foote, C. S., Dzakpasu, A. A. & Lin, J. W.-P., *Tetrahedron Lett.*, 1975, 1247.
11. Ando, S. & Kabe, Y., *J.C.S. Chem. Commun.*, 1984, 741.
12. Laarhoven, W. H., *Rec. Trav. Chim. Pays-Bas*, 1983, **102**, 241.
13. Scaiano, J. C., *J. Photochem.*, 1973, **2**, 81.
14. Bryce-Smith, D., ed., *Photochemistry – a Specialist Periodical Report*, The Chemical Society, Vols. 1–14.
15. Cohen, S. G., Parola, A. & Parsons, G. H., *Chem. Rev.*, 1973, **73**, 141.
16. Bruce, J. M., in *The Chemistry of Quinonoid Compounds*, ed. Patai, S., Wiley Interscience, Pt. I, p. 465.
17. Foster, T., Elliot, A. J., Adeleke, B. B. & Wan, J. K. S., *Canad. J. Chem.*, 1978, **56**, 869.
18. Akiba, M., Kosugi, Y. & Takada, T., *J. Org. Chem.*, 1978, **43**, 4472; Falci, K. J., Franck, R. W. & Smith, G. P., *ibid.*, 1977, **42**, 3317.
19. Levy, N. & Cohen, M. D., *J.C.S., Perkin Trans. II*, 1979, 553; Wollenben, J. & Testa, A. C., *J. Phys. Chem.*, 1977, **81**, 429.
20. Havinga, E. & Cornelisse, J., *Pure and Appl. Chem.*, 1976, **47**, 1; Cornelisse, J., de Grunst, G. P. & Havinga, E., *Chem. Rev.*, 1975, **75**, 353; Cornelisse, J., Lodder, G. & Havinga, E., *Rev. Chem. Intermed.*, 1974, **2**, 231.
21. Van Riel, H. C. H. A., Lodder, G. & Havinga, E., *J. Am. Chem. Soc.*, 1981, **103**, 7257.
22. Den Heijer, J., Shadid, O. B., Cornelisse, J. & Havinga, E., *Tetrahedron*, 1977, **33**, 779.
23. Fischer, M., *Angew. Chem. Internat. Edn.*, 1978, **17**, 16.
24. Barton, D. H. R., *Pure and Appl. Chem.*, 1968, **16**, 1.
25. Hesse, R. H., *Adv. Free Radicals*, 1969, **3**, 83.
26. Coombes, L. G. & Sutherland, I. O., eds., *Comprehensive Organic Chemistry*, Pergamon, 1979, Vol. 2, p. 356.
27. Christl, M. & Brunn, E., *Angew. Chem. Internat. Edn.*, 1981, **20**, 468.
28. Wilt, J. W. & Niinemae, R., *J. Org. Chem.*, 1980, **45,**, 5402.
29. Berson, J. *et al.*, *J. Am. Chem. Soc.*, 1982, **104**, 2217; 1981, **103**, 684; 1980, **102**, 3870; *Acc. Chem. Res.*, 1978, **11**, 446.

30. Moss, R. A., Guo, W., Denney, D. Z., Houk, K. N. & Rondan, N. G., *J. Am. Chem. Soc.*, 1981, **103**, 6164.
31. Lawson, G., Moody, C. J. & Pearson, C. J., *J.C.S. Chem. Commun.*, 1984, 754.
32. Zimmerman, H. E. & Hovey, M. C., *J. Org. Chem.*, 1979, **44**, 2331.

INDEX